D0855944

ROBERT E. KRIEGER
PUBLISHING COMPANY INC.
MALABAR
FLORIDA 32950

FLUID LOGIC CONTROLS AND INDUSTRIAL AUTOMATION

FLUID LOGIC CONTROLS AND INDUSTRIAL AUTOMATION

DANIEL BOUTEILLE

with the cooperation of

Claude Guidot

Translated, revised, and edited by
Stuart North and Leonard P. Gau
with the cooperation of Daniel Bouteille

A WILEY-INTERSCIENCE PUBLICATION

JOHN WILEY & SONS, New York • London • Sydney • Toronto

Library of Congress Cataloging in Publication Data:

Bouteille, Daniel.
Fluid logic controls and industrial automation.

"A Wiley-Interscience publication."
Original ed. published in 1970 under title: Les commandes logiques à fluides et l'automatisation industrielle.
1. Fluidic devices. 2. Automatic control.
I. Title.

TJ853.B6813 629.8'042 73-520
ISBN 0-471-09172-3

Printed in the United States of America

10 9 8 7 6 5 4 3 2 1

FOREWORD

I am happy to have been a part of the fortunate series of circumstances which led to cooperation by United States and French publishers, their editors, and the author who organized this collective work.

The technology of fluid logic is both new and exciting, and a world-oriented book will be welcome by engineers at all levels, especially those who like straightforward explanations of what fluid logic is all about.

Over 100 companies have already entered the field of all-air control, often called "fluidics" or "pneumatic logic," and hundreds more will follow. Control engineers will have more choices now that pneumatic control is available to do jobs previously reserved for electromechanical or electronic control. Thousands of installations worldwide have proved the principles. Unifying standards which will give the pneumatic logic circuit designer a valuable tool are now being written.

The author of this book, Daniel Bouteille, a French engineer, and its editors, Stuart North and Leonard Gau, both United States specialists in the field, are among the many technical pioneers who are advancing the art and science of pneumatics. They are all uniquely qualified to "tell it like it is."

FRANK YEAPLE
ASSOCIATE EDITOR
Product Engineering Magazine

July 1973

PREFACE

As the editors, it has been our privilege to assist the author in the transformation of this important work from its original French version. The popularity of this book in Europe has resulted in this new edition, which has been revised and expanded to include the developments in the United States, as well as the latest contributions of all industrial countries. The worldwide interest in fluid logic for industrial control applications will assure its universal acceptance as one of the major control technologies. This book is intended as a practical guide to the state-of-the-art, with emphasis on industrial applications.

Students and experienced control system designers will benefit from the authoritative discussion of the basic operating principles presented in a clear, concise manner and supported by complete illustrations, depicting the operation and required combinations of devices to implement the basic control functions.

Many charts and graphs are included to assist the reader in making direct comparisons between the operating characteristics of the various logic and peripheral devices.

Complete chapters have been devoted to each of the following major topics: movable and nonmovable part logic devices, peripheral devices, selection of a control technology, a simplified method of circuit design, and typical industrial applications.

Since a wide variety of graphic symbology systems are popular throughout the world, no attempt has been made to arbitrarily recommend a particular one to the reader. Instead the author has employed one of the systems popular in Western Europe.

STUART NORTH
LEONARD P. GAU

July 1973
Rochester, New York
Birmingham, Michigan

CONTENTS

FLUID LOGIC CONTROLS AND INDUSTRIAL AUTOMATION

INTRODUCTION

In the midst of the tremendous advances of electronic and electrical controls, the ancient art of pneumatic and hydraulic power has provided the basis for the new technology of fluid logic controls.

The historical significance of fluid power has fore-shadowed the important role that fluid logic controls now have in satisfying the worldwide demand for industrial automation. The multiplicity and diversity of international technical developments within the last decade have created the need for a handbook that bridges geographical borders and professional barriers in documenting the methods available to facilitate the utilization of fluid logic controls.

This book describes and classifies the hardware and software developed by the major industrial countries to meet the requirements of advanced fluid logic control systems. The similarities between fluid logic controls and electrical control technologies are cited to assist those familiar with the latter in acquiring a rapid understanding and appreciation of the differences and unique advantages of this new control art.

This book is designed for those interested in acquiring a practical knowledge of the subject, or for individuals already involved in the field and desiring to increase their appreciation of fluid logic by learning of the developments in all of the countries presently involved in advancing the technology. The scope of this book is defined and delineated by the title. This indicates that the technology of fluid analog controls for proportional instrumentation has not been included. Also the practical application of fluid logic controls has been limited to industry, thereby excluding the medical and military fields, space exploration, and such.

CHAPTER 1

INDUSTRIAL AUTOMATION: THE NEED FOR FLUIDS

1-1 GENERAL PRINCIPLES OF AUTOMATION

1-1-1 Introduction

The automation of industry can manifest itself in a variety of forms, depending on the type of industry and its production requirements.

We shall begin by suggesting that the primary aim of automation is to first help, then perhaps eventually replace men who work at tiring or repetitive tasks. This new mechanization helps obtain economic production with more uniform quality.

The production process, and the design of products, is often modified by automation. The consequences of introducing automatic production then go far beyond simple aids to, or the replacement of working men.

However, to simplify this presentation, we shall use that comparison between the working man and the machine designed to help, or replace him, as the basis for our study examples.

1-1-2 The Stages of Automation

The first stage in the process is to replace muscle power with *mechanical power,* which is controlled directly by a worker. The worker can then apply his energies to increasing productivity and production efficiency. It is a fact that power presses have replaced manual hammer work, thus providing greater mechanical power, and have fostered production, uniform quality, and the reduction of worker fatigue.

The second stage is to coordinate automatically the movements of various power *devices* with *control systems.* This method still requires the worker to feed manually and remove the component parts and to check the production quality.

The final stage is the automation of component part feeding and removal, along with automatic quality inspection. At this stage the machine is completely automated—the workers mental and physical energies are no longer directly required in the manufacturing process. The worker will be needed only to monitor the machine's operation and to make adjustments or repairs when there is a failure.

1-1-3 The "Control Power" Dialogue

Figure 1-1 illustrates schematically the general organization of an automated machine. A typical machine consists of the following:

The Power System. In the example it is a drill head mounted on a cylinder to provide feeding action.

The Control System. This coordinates the automatic cycle of the machines power system and allows for manual intervention by the operator. The control system also includes:

- The input devices, consisting of the required sensing devices.
- The logic circuit, which contains the program instructions.
- The manual control panel for operator intervention.

To ensure proper operation and to protect the machine from damaging itself or the operator when a malfunction occurs, a complete dialogue is usually maintained between the power system and the control system.

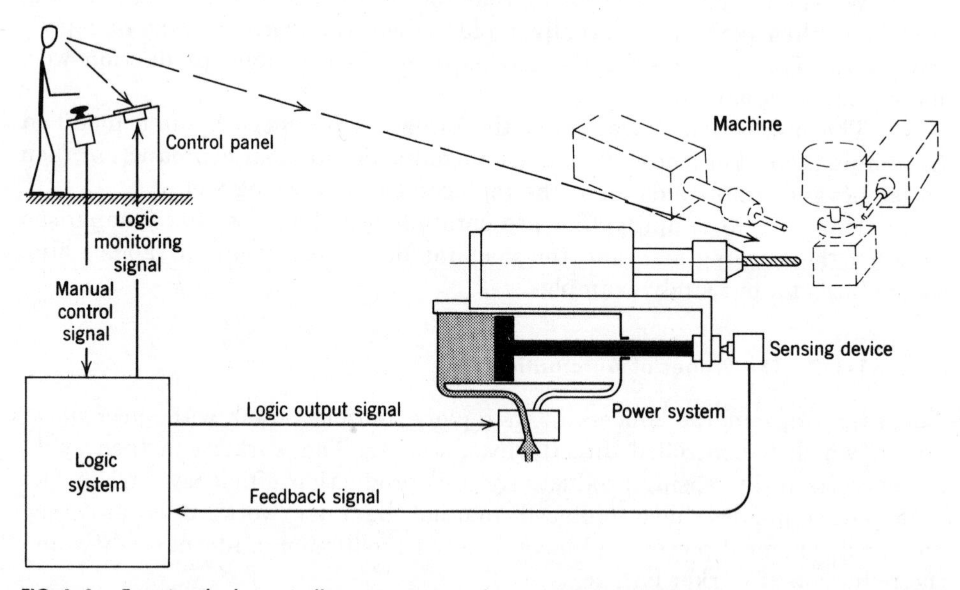

FIG. 1–1. Functional schematic illustrating a typical automation control loop.

1-1-4 Analog and Digital Controls

According to the type of controls used, the control-power dialogue takes on various forms. *Analog controllers* respond to all signal values between two extremes. Many regulating-type controllers (pneumatic and electric) employ analog control techniques.

Many industrial control systems employ *digital controls.* They respond only to the two extremes—"full on" or "full off." Some examples of this condition are: an electric wire is conducting electricity ("on") or not conducting ("off"); a pipe or tube filled with a fluid under pressure ("on") or open to discharge or exhaust path ("off"); an electric relay's contacts closed ("on") or open ("off"). Control devices that respond to "ON–OFF" signals are known as digital controls.

Only the fully extended or fully retracted positions of the drill and feed cylinder combination in Fig. 1-1 are of importance to a digital controller. (As explained in the introduction, digital controls are the only ones described in this book.)

The digital "control power" dialogue is simplified to the point that in most applications the input devices or sensors are actuated only at the beginning or conclusion of an action or movement. Only at these two points does the control system make a decision based on the feedback information to the inputs. The decision is based on the preprogrammed instructions.

1-2 THE NEED FOR FLUID AUTOMATION

1-2-1 Fluid Power Systems

The sources of energy usually employed to power automatic machines are hydraulic or electric motors, and steam or combustion engines which usually produce a rotating motion. Man has endeavored to employ a rotating movement to accomplish what he previously had to do by hand. A typical example of this is the replacement of the hand file with the rotating motion of machine tools—lathes, grinders, and milling machines.

There are two methods of transforming a rotating motion into a linear movement:

Mechanical Methods. Cams, connecting rods, nuts and screws, and such.

Fluid Power Methods. These utilize cylinders that are filled with fluid under pressure, produced by a rotating pump or compressor.

The fluid power method has the ability to transmit the pressurized fluid to a number of locations where it is needed, rather than transforming rotary motion at each location. Aside from transforming rotating motion to linear movement, there is a need for the transmission of this converted energy.

Mechanical transmission devices have made possible many important industrial machines. A typical example is the screw machine which extensively utilizes the mechanical transmission power from one rotating shaft to a variety of linear movements in various locations, with the assistance of levers, push rods, cables, and such.

Later, *electrical transmissions* demonstrated their advantages over the geometric design requirements of mechanical transmission. Electrical devices provided easily applied methods of production, fast rotating motion (electric motors) or short linear movements (solenoids.) Long linear strokes are not possible without mechanical transformation. The so-called electric cylinders consist of an electric motor, speed reduction gears, and a nut and screw system to produce the linear movement.

Finally, the development of *fluid power transmission* provided the ease of application of electrical transmission and the simplicity of mechanical transmission.

Pressurized fluids are distributed through pipes, hoses, and tubes to each point of use. The most common power systems are cylinders that permit all types of linear movements.

Two complementary techniques within the technology of fluid power are hydraulics and pneumatics.

Hydraulic power utilizes noncompressible fluids like oil or water, permitting a standard pressure range, 1500–4000 psig, which produces equivalent linear forces from cylinders with accurate speed control. Basically each machine requires an electric motor, a pump, and a fluid resevoir.

Pneumatic power utilizes compressed air. Standard pressure range, 40–150 psig, limits the thrust obtainable from air cylinders. Air-cylinder speed cannot be accurately controlled due to the compressibility of air, except with the use of air over oil systems.

Compressed air is easily distributed in a factory from a central compressor and reserve tank through overhead piping manifolds to any point where it is required. Spent air is exhausted harmlessly into the factory atmosphere.

The use of pneumatic power in industry has developed rapidly during the last quarter century. Its ease of application has made possible the substitution of machine power for man power in all areas where it has been applied. Only when large mechanical forces are required on big power presses or for handling of heavy machinery is hydraulic power substituted for pneumatics. The only exception is when highly accurate speed control is required on applications where air over oil is not practiced.

The next section answers the question regarding the need and importance of fluid logic controls in industry.

1-2-2 Fluid Logic Controls

Two principal reasons have led to the development of fluid logic controls: the extensive use of fluid power and the need to work in explosive atmospheres.

Originally, pneumatic analog control techniques were created to replace electrical controls in the petroleum and chemical industries, where there was an explosive atmosphere. Later in this book we describe how certain fluid logic systems, using the same operating pressures as analog controls, became very complementary and were first accepted within these industries.

The development of fluid logic controls has followed a natural course so that today they are used primarily to control fluid power systems. When pneumatic or hydraulic cylinders are used to replace a worker's movements, there is usually no need for analog controls. The cylinder's movements are usually monitored by limit valves or sensors which are actuated at the fully extended and fully retracted positions. This arrangement is ideal for digital controls which can easily coordinate the control and feedback signals.

Before the development of fluid logic controls, electrical logic (relays) were the primary means of controlling fluid power systems. These techniques are called Electro-pneumatic or Electro-hydraulic systems, depending on which media is employed.

The development of totally pneumatic systems made it obvious that the homogeneity of devices from inputs to outputs had many advantages over electrical controls. These include no electrical devices to burn or short out greater reliability; longer life for the controls; elimination of solenoid valves; immunity to mechanical shock, vibration, and contaminated environments; no special enclosures required in explosive environments; safety for operators and maintenance people. Other advantages include the requirement of only one trade union for adjustment or service; the superiority of pneumatic sensors; the case of understanding the technology; the simplicity of the components design; the speed of operation; the minimum maintenance requirements, and such.

After a discussion of the various types of control devices covered in Chapter 3, peripheral equipment, input devices, interfaces, manual controls, and such are described in Chapter 4. In Chapter 5 we make a comparison between pneumatic, electric, and electronic logic devices, and also mention the preferential fields of application for fluid logic in industry.

CHAPTER 2

LOGIC FUNCTIONS AND INDUSTRIAL AUTOMATION

2-1 THE FUNDAMENTALS OF DIGITAL LOGIC CIRCUITS

It is a very common practice to design digital automation circuits by direct thought. Electricians frequently connect relays in series in cut and try arrangements until they satisfy the circuit requirements.

This circuit design method requires specialists in a given technology to ensure success. For example, the ladder diagram method used by electricians employing relays is far different from the "cascade" method used by pneumaticians employing double four-way valves.

Nonlatching relays require an input "feedback" from the relay output to maintain the relay in the actuated position.

Double pilot pneumatic spool valves do not require feedback; they maintain the shifted position by friction.

To expedite the design procedure and minimize the possibility of errors, it is advisable to systemize circuit design. The best tool for this purpose is binary algebra, which considers only two values—1 and 0. Digital controls also consider only two values; this is one application of binary algebra. When binary algebra was first applied to electrical circuits, it was known as contact algebra. Eventually it was named Boolean algebra after the English mathematician George Boole who defined it during the last century.

The logic functions are the basic operations of Boolean algebra. These functions are also the basis for the creation of electronic and fluid logic devices. The basic logic functions are important in the practical application of these techniques.

The advantages of using Boolean algebra follow:

- The resulting circuits are practical, easily designed, readily understood, and easy to trouble-shoot regardless of their complexity.
- The control modules are identified by the logic functions to which they correspond.

- Education of engineers and technicians in the techniques of design, construction, installation and maintenance, and trouble-shooting is easier. A basic universal language exists for all control technologies: electrical, electronics, pneumatics, and fluidics.

Therefore the conversion from one to another is greatly simplified and requires only a minimum of time for one to become familiar with differences of circuit symbols and hardware.

Many technical schools and colleges now include in their curriculum courses in Boolean algebra and control logic, oriented to the needs of engineers and technicians dealing with automation.

In this chapter we discuss the basic background required in understanding logic function and control circuits. In Chapter 3, this basic information will be applied to the technology of fluid logic and to the design of the devices that are needed to create the logic functions.

In Chapter 6, Boolean algebra is discussed in more detail. Chapter 7 follows with a thorough explanation of a logical method of solving automation control problems. In Chapter 8, a collection of typical industrial automation problems is discussed as an aid in applying the techniques described in this book.

2-2 LOGIC AND FLUID CONTROLS

In digital control logic, as derived from Boolean algebra principles, only binary logic devices can be used, that is, only devices that can create two discrete states: 0 or 1.

In the technology of electrical controls these two signal states consist of the presence or absence of voltages. In the technology of fluid controls these two signal states take the form of air compressed to levels higher than the atmospheric pressure for the 1 state; and compressed air exhausted back to the atmospheric pressure for the 0 state.

Fluid logic functions are always related to the binary values of the pressures acting on the various devices within the control circuit. Each fluid logic device is designed to respond to specific logic functions.

2-3 THE BASIC LOGIC FUNCTIONS

Figure 2-1 illustrates the basic logic functions, YES, NOT, OR, AND, and their symbols. These functions are the basic control methods.

The YES function relates to an equality of states; for example, if a device provides an output pressure when a signal is applied to its input port, an equality of states exists between the output signal U and the input signal a. The inverse of this situation is also applicable. The device creates the YES function:

$$\boxed{U = a} \quad \text{if} \quad U = 1 \quad \text{when} \quad a = 1$$

$$\text{and} \quad U = 0 \quad \text{when} \quad a = 0$$

The NOT function results from an inversion of the state: if a device provides an output pressure when an input signal is absent, we have an inversion at the output U of the input signal a. The device creates the NOT function:

$$\boxed{U = \bar{a}} \quad \text{if} \quad U = 1 \quad \text{when} \quad a = 0$$
$$\text{and} \quad U = 0 \quad \text{when} \quad a = 1$$

The OR function (two input type) combines the states of two signals a and b. An OR device provides an output at port U if one OR the other (or both) of the input signals are "on" (state 1).

$$\boxed{U = a \text{ OR } b} \quad \text{if} \quad U = 1 \quad \text{when} \quad a \text{ OR } b = 1 \quad \text{(or both)}$$
$$\text{and} \quad U = 0 \quad \text{when} \quad a \text{ and } b = 0$$

The AND function (2 input type) combines the states of two signals a AND b. An AND device creates an output at port U when input a AND b are both "on" (state 1).

$$\boxed{U = a \text{ AND } b} \quad \text{if} \quad U = 1 \quad \text{when} \quad a \text{ AND } b = 1$$
$$\text{and} \quad U = 0 \quad \text{when} \quad a \quad \text{or} \quad b = 0 \quad \text{(or both)}$$

The symbology is simplified in Boolean algebra; $U = a$ OR b is written $U = a + b \longrightarrow$logic sum, and $U = a$ AND b is written $U = a \cdot b \longrightarrow$logic product. Figure 2-1 illustrates the symbols for each of these functions that we shall use when drawing circuits.

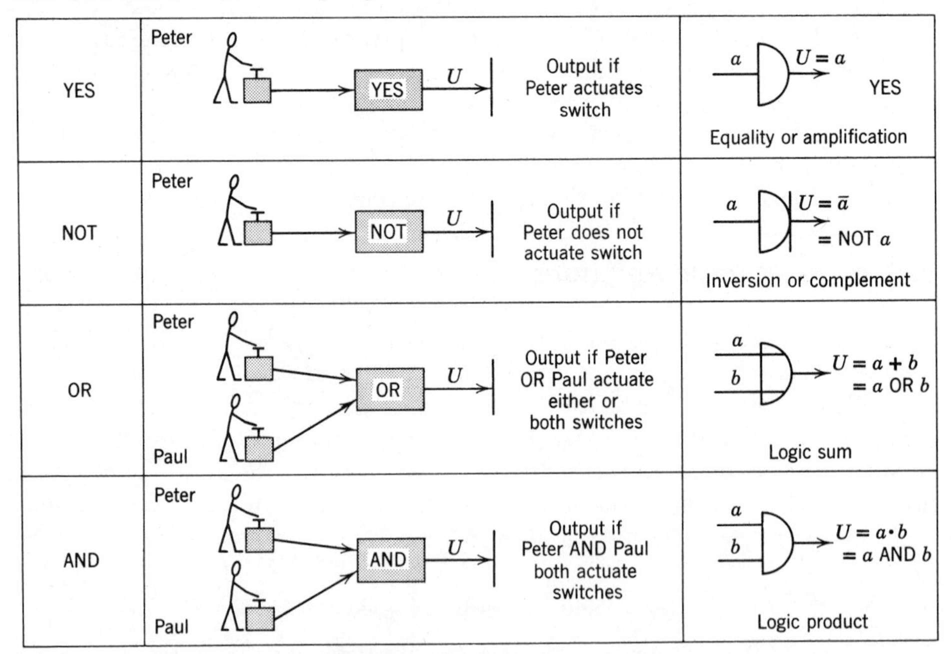

FIG. 2–1. Basic logic functions—definitions and symbols.

2-4 A SIMPLE APPLICATION EXAMPLE

Many problems can be solved with use of the basic functions described in Figure 2-2.

If the press cylinder A is lifted to a, and if NO object is present at position b, and if the operator protection shield is down in position e, it is then possible to create the control command from c OR d to extend cylinder B which will position a part to b. This can be converted into Boolean terminology:

$B1$ receives a control command if we have

$$a \text{ AND NOT } b \text{ AND } e \text{ AND } c \text{ OR } d$$

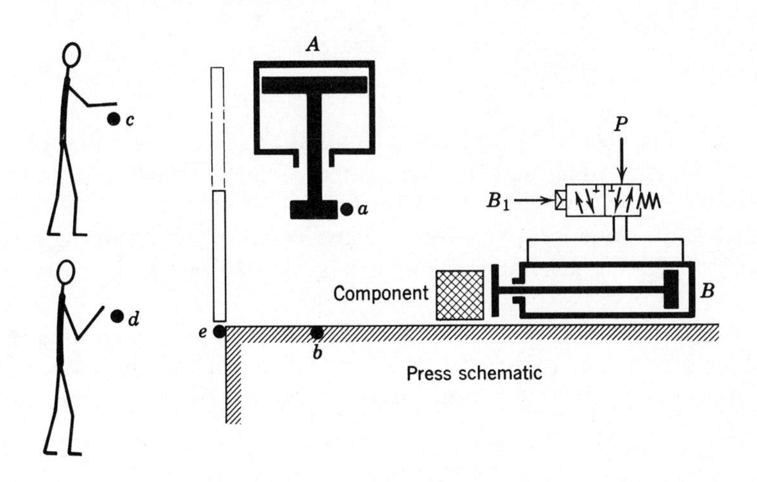

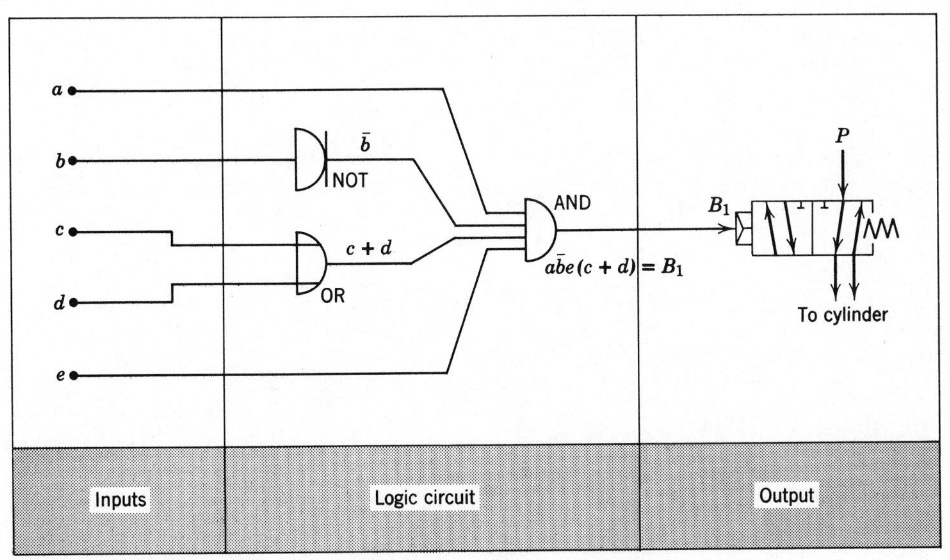

FIG. 2-2. Illustrating a press schematic and its control circuit diagram.

This can be expressed by the equation:

$$B_1 = a \cdot \bar{b} \cdot e \cdot (c + d)$$

AND NOT AND AND OR

The circuit in Fig. 2-2 resolves the problem—*a, b, c, d,* and *e,* when actuated, provide inputs at the logic circuit, which, if the required conditions are satisfied, will provide an input to pilot port *B*1, which actuates the control valve operating cylinder *B* which positions a part.

The logic function circuit consists of the following:

- A NOT function to obtain $\bar{b}$.
- A two-input OR function to obtain $c + d$.
- A four-input AND function combining all factors of the logic product.

2-5 THE AUXILIARY FUNCTIONS

To solve all automation control problems, two auxiliary functions are required in addition to the four functions just described. The MEMORY (flip-flop) function and the TIMING function are described in Fig. 2-3.

The MEMORY function provides a means of storing or retaining a specific signal for subsequent use. When a signal is applied to the *memory* it is memorized.

The TIMING function provides a means of delaying a signal for a specific duration before it is available for use. It is not a Boolean logic function; however, it satisfies an important requirement in many automation circuits.

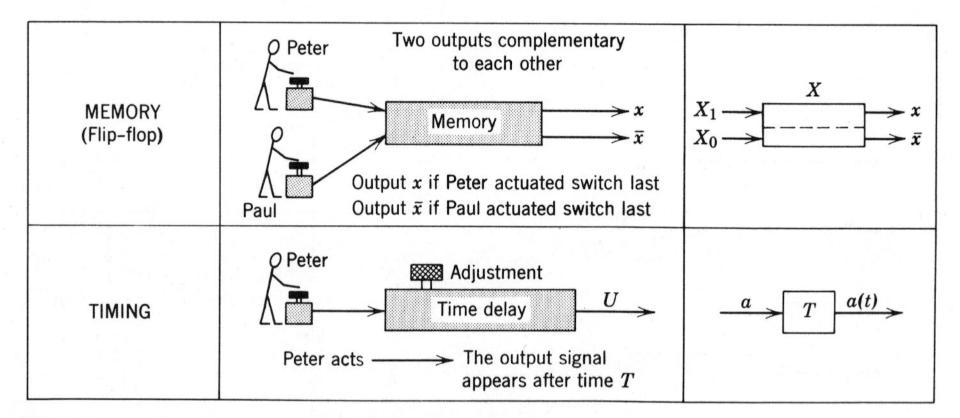

FIG. 2-3. Auxiliary functions—definitions and symbols.

2-6 THE CONSTITUENTS OF A LOGIC SYSTEM

A complete logic system includes all the devices within a specific technology (electrical, electronic, pneumatic, or fluidic) that are necessary to solve a particular control problem.

There are several ways to construct a complete logic system. The first option is to employ individual devices for each logic function—YES, NOT, OR, AND, and the auxiliary functions, MEMORY and TIMING.

Another option is to construct a system that employs one basic "universal" logic function device which creates all the functions previously described. The universal logic functions are described below.

A universal logic function is capable of creating all of the basic logic functions, when duplicates of itself are interconnected into specific configurations. Figure 2-4 illustrates how these functions can be created with only NOR functions or only with INHIBITION functions—the NOR function is universal as is the INHIBITION function.

The NOR function or NOT-OR function, $\overline{a + b}$ is the inverse of an OR function $a + b$; hence it uses the same symbols, except that a bar is drawn over the expression. This function is common in electronics. Later we describe fluid logic control systems employing only NOR functions.

The INHIBITION function, $\bar{a} \cdot b$ is the logic product of two signals, one of which is inverted. It employs the same symbols as the AND function with the exception of a bar over one of the letters. Some of the fluid logic control systems that we describe later employ only one universal INHIBITION function device.

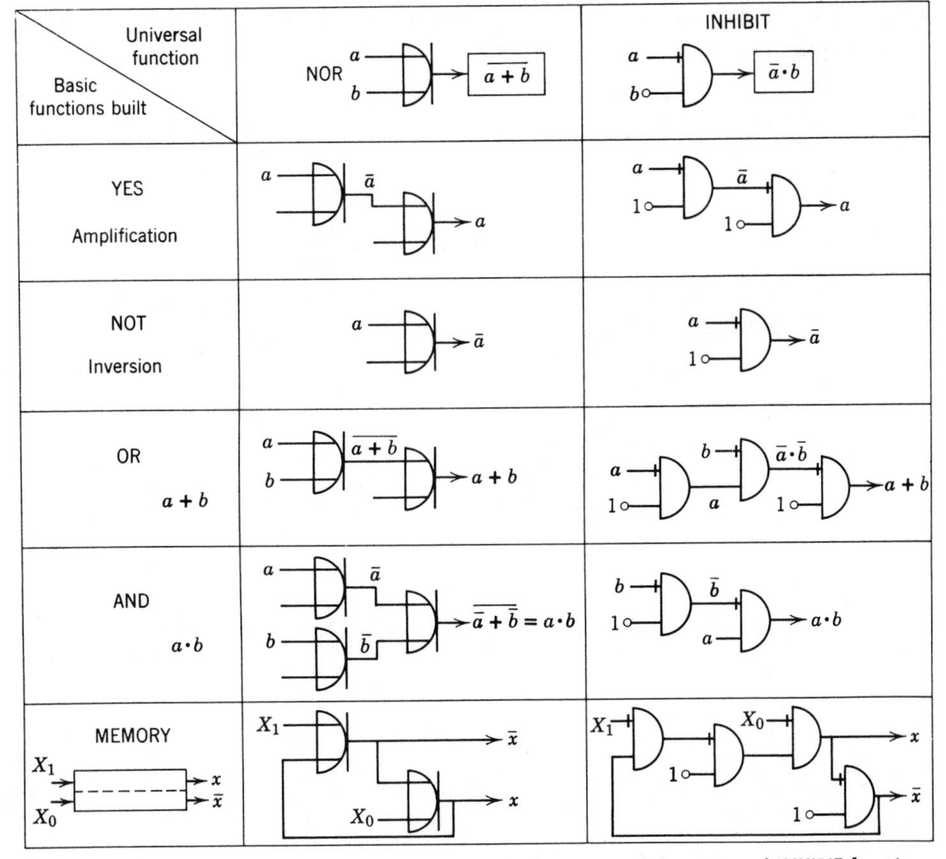

FIG. 2-4. Logic functions built from: the universal NOR function and the universal INHIBIT function.

CHAPTER 3

FLUID LOGIC DEVICES

3-1 DEFINITION

We have described how devices are selected and assembled into control systems for automating production machinery. Selecting the most appropriate technologies—electrical, electronic, hydraulic, pneumatic, or fluidic—depends on the application and its requirements.

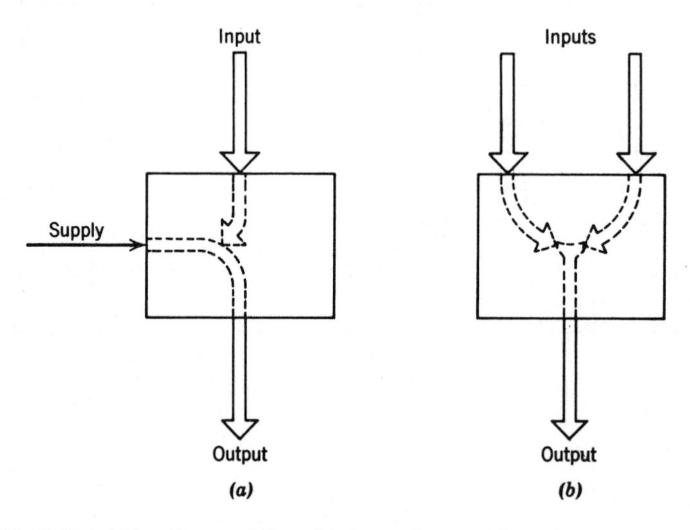

FIG. 3–1. The distinguishing characteristics of active and passive logic devices. (*a*) Active device; (*b*) passive device.

Each technology has the devices necessary for creating the required control circuitry. Devices designed for digital control are logic devices. Figure 3-1 illustrates the differences between the two basic types of logic devices.

Active logic devices divert a maintained supply to the output port when the required input conditions are satisfied. There is an amplification of the input signal as a result of the higher level supply signal which is maintained.

This type of device is often called a relay or amplifier, because it propogates, relays, or amplifies the input signal. The power amplification is expressed by the ability to "fan out" or divide the output signal into a number of control signals, useable as inputs to other identical or similar devices.

Passive logic devices create an output signal directly from the input signals; there is no amplification of the input signals. No maintained (active) supply is utilized, and more than one input port is required to create a logic function.

3-2 TECHNOLOGICAL CLASSIFICATIONS OF FLUID LOGIC DEVICES

There are many types of fluid logic devices currently available. For the purpose of our discussion, we limit ourselves to those suitable for industrial applications. Even within this somewhat limited context, there are a great variety of principles and phenomena that do not always fit neatly into specific classifications.

Some devices are suitable for operation with liquids—usually water or oil. The use of gas is preferred, primarily because of the smaller pressure losses that result from their use in tubing or piping. Gases such as nitrogen, oxygen, natural-gas, and carbon dioxide have worked well in specific applications. A good example of this application is the safety stations built into long natural-gas pipe lines.

Except for special applications, such as the one mentioned above, compressed air is universally employed as the work medium.

The two main reasons for this use are that compressed air supplies are readily available in most factories, and that the exhausting of used air to atmosphere is safe and simplifies the interconnections between devices.

A more valuable classification concerns the operating pressures and the diameters of the passages of the various devices. Both of these criteria are used to determine the efficiency of the devices when combined into control circuits. These important factors follow:

- The operating pressure.
- The flow, related to the operating pressure and passage diameter.
- The power, resulting from the pressure and flow.

Figure 3-2 distinguishes between the four basic types of fluid logic devices, based on their operating pressures and the diameter of their flow passages.

The nonmovable-part devices include:

- The jet destruction devices comprising the laminar/turbulent flow devices (or turbulence amplifiers) and the impact modulators.
- The jet deflection devices, which include those using the following principles: jet interaction, wall attachement, and vortex feedback.

The movable-part devices include:

- The low pressure movable-part devices designed to operate on pressure within the range of 3 to 30 psig.
- The high pressure movable-part devices designed to operate on industrial air pressures within the range of 30 to 120 psig.

Each type of device will be described as to operating principles, design requirements, and applications in circuits.

In Chapter 5 we use information in Fig. 3-2 to provide a practical comparison between each type of logic device system.

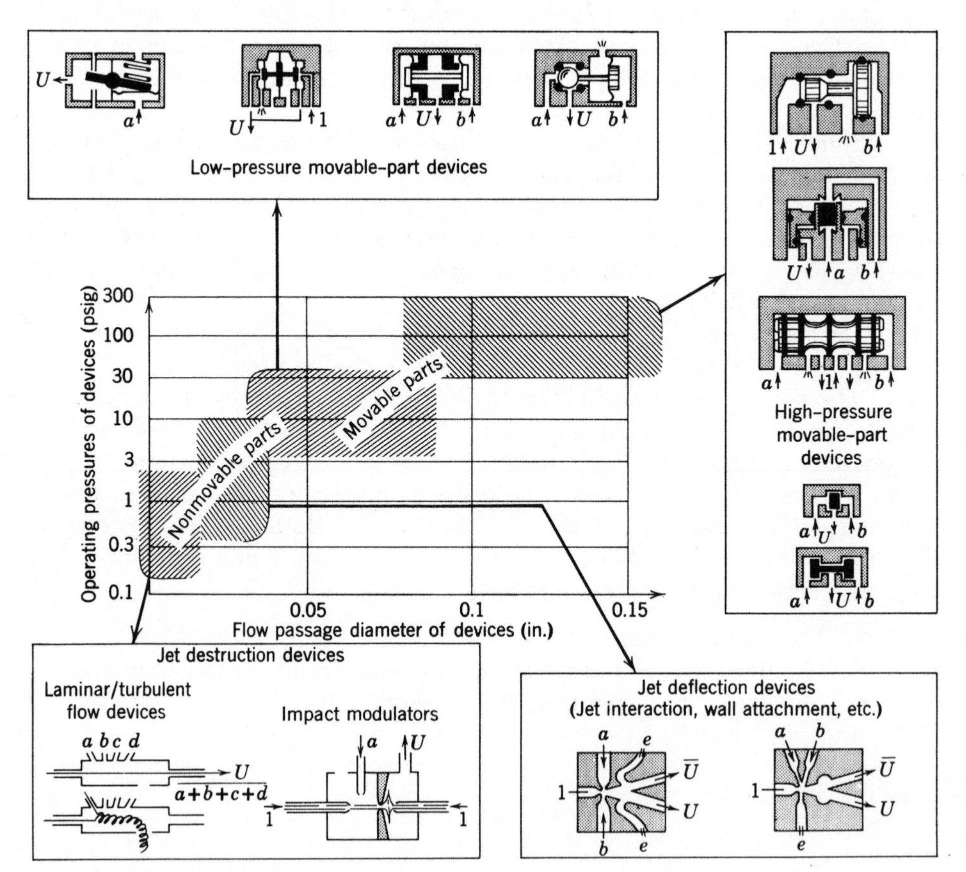

FIG. 3-2. Technological classification of fluid logic devices.

3-3- NONMOVABLE-PART DEVICES

3-3-1 Characteristics Common to All NonMovable-Part Devices

The word "fluidics" is usually used in reference to nonmovable-part devices. The word originated in the United States and is now commonly accepted internationally. Due to the nonmovable-part design, these devices theoretically have a potentially unlimited life expectancy.

Fluid switching is obtained by jet interaction. In active devices, the supply is a maintained flow that is interrupted by the action of the input control signals which utilize various phenomena. As a result of the requirement for a maintained supply, the air consumption is continuous in both operating states 1 or 0.

These devices operate on low pressure generally less than 10 psig. The two main reasons for this low pressure operating range follow:

1. Specific flow interaction phenomena cannot function at higher pressures.
2. The rate of air consumption at high pressures would make these devices impractical.

3-3-2 Jet Destruction Logic Devices (Laminar/Turbulent Flow Devices and Impact Modulators)

3-3-2-1 History

R. N. Auger, an American electronics engineer, was the first to apply the principles of laminar/turbulent flows in the early 1960s in cooperation with the United States company Howie Corp. (now Agastat).

The development of wall attachment devices preceded the application of laminar/turbulent flow devices in the United States. Therefore, the adoption of laminar/turbulent flow devices was more rapid in Europe at first.

Since 1965, the British company Maxam has manufactured devices based on principles defined by Auger. Maxam also designed the peripheral devices necessary to provide complete industrial systems.

Shortly after Maxam, the Technical University of Holland in Delft designed the first planar laminar/turbulent flow device, a flat device easily manufactured by injection moulding.

In the United States Pitney Bowes Fluidics Controls (now Asco Fluidics) the fluid division of the Bailey Meter Company and Fluidic Industries Inc. have developed planar type laminar/turbulent flow devices, both constructed from injection molded plastic. These United States manufacturers and others in Germany, Holland, France, Czeckoslovakia, and the USSR have contributed to the proliferation of laminar/turbulent flow devices.

The impact modulators have been developed by Johnson Service, a manufacturer in the United States.

3-3-2-2 Operating Principles of Laminar/Turbulent Flow Logic Devices

Figure 3-3 illustrates the "active" operating principles employed in laminar/turbulent flow devices. The very low pressure supply creates a laminar flow. This type of flow has the ability to travel through the atmosphere of the device interaction chamber and exit from the output port.

The laminar flow to output continues until an input signal enters the chamber and interacts with the supply flow. This interaction destroys the laminar flow, which in turn stops the supply from reaching the output port. In this state the supply flow is vented to atmosphere.

Figure 3-4 illustrates the operating characteristics of a particular type of laminar/turbulent flow device.

A demonstration of jet destruction of laminar flow. (Courtesy of Pneumotech AG, Switzerland.) This experiment conducted by the Swiss engineer, G. Glaettli, shows how stained water was used to illustrate the principle of laminar/turbulent flow operation. The top view emphasizes how well the coupling stream reaches the output collector on the right. Note that very slight waves are developing in the stream at this point. The lower view shows how a control signal intercepts the coupling stream and destroys it through turbulences without imparting sufficient energy to deflect the stream, thereby demonstrating very great sensitivity of this principle.

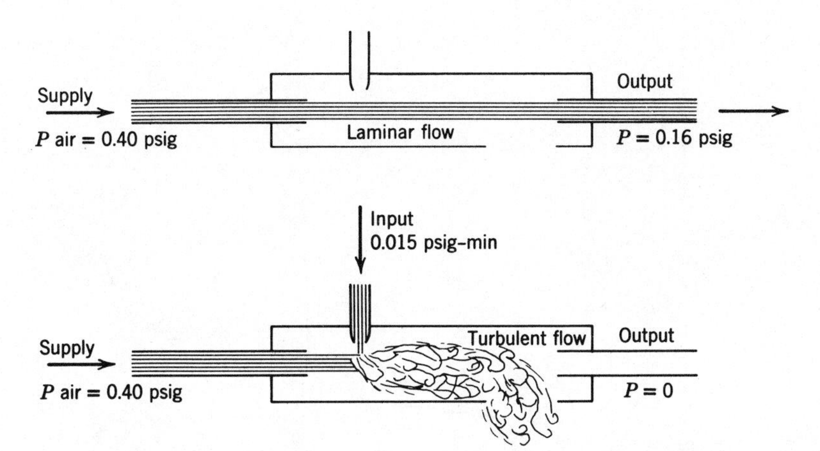

FIG. 3–3. The operating principle of laminar/turbulent flow devices.

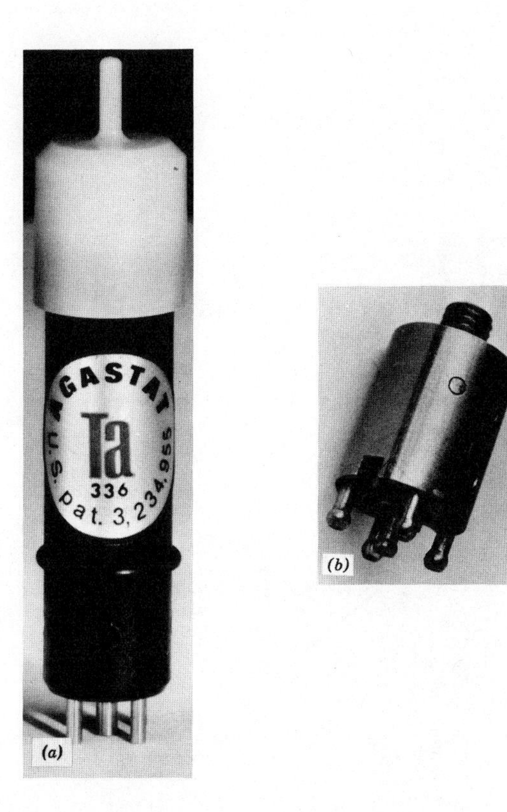

Laminar/turbulent flow devices. (Courtesy Agastat Division, Amerace Esna Corp., USA; and the Institute of Control Sciences, USSR.) Both of these devices operate according to the principles of laminar/turbulent flow as described in Fig. 3–3. Each is designed to have air supplied to the single port end. Device (a) from the USA plugs in while device (b) from the USSR screws in. Both have single air collectors at the opposite end, surrounded by four control port fittings.

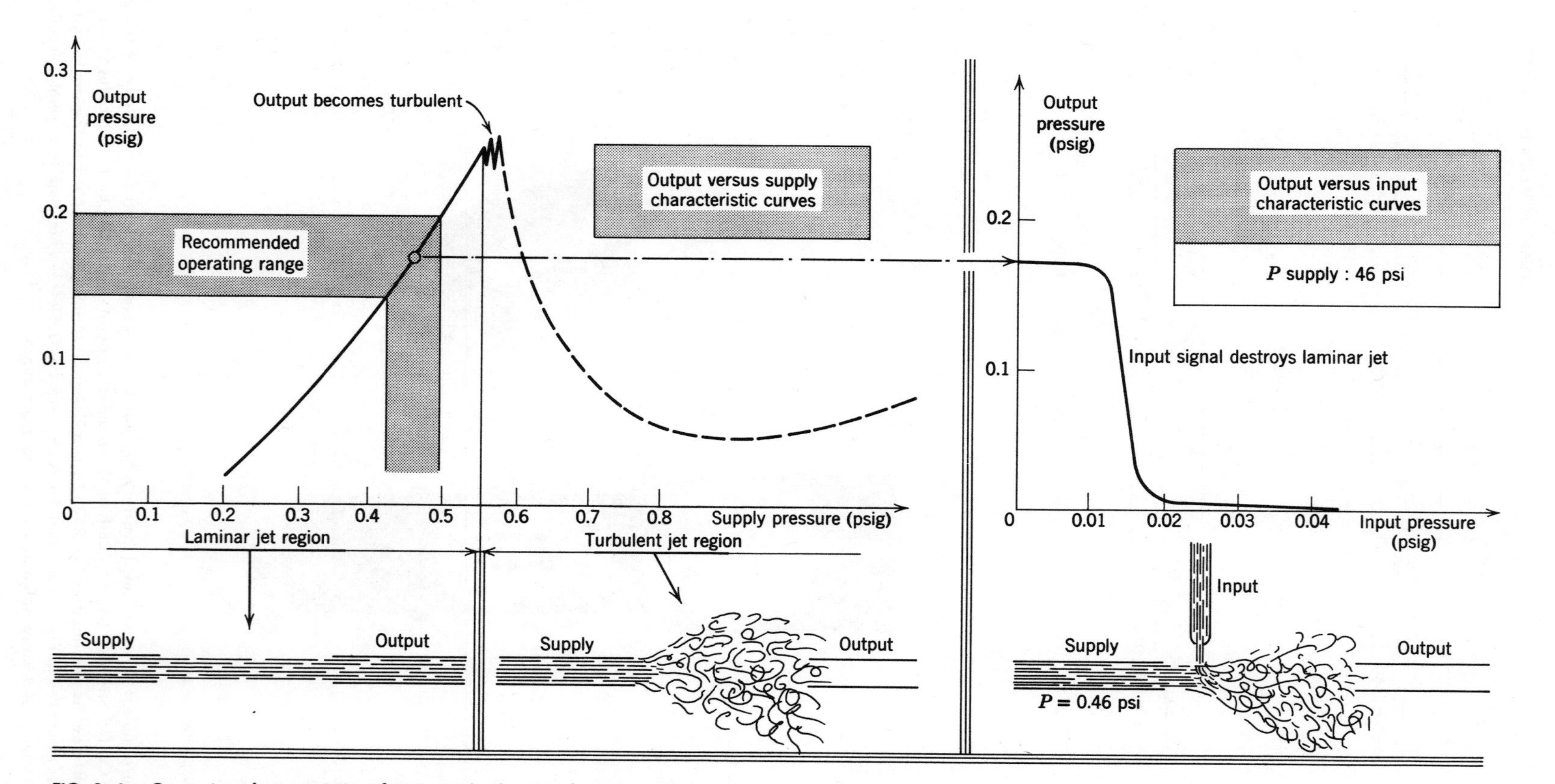

FIG. 3–4. Operating characteristics of a particular laminar/turbulent flow device.

The Output Pressure versus Supply Pressure Curve. This curve illustrates the limited range in which laminar flow can be maintained. Only a low Reynolds number will allow a laminar flow. The formula is

$$R = \frac{V^2}{\rho}$$

where

V = the velocity of flow

ρ = the specific weight of the fluid

The ρ for air is small when compared to the ρ for water; therefore, only a low velocity is obtained by a low supply pressure in the supply tube.

As illustrated by the curve in Fig. 3-4, the laminar flow is only possible with a supply pressure under 0.66 psig for this particular device. In practical application this device is used with supply pressure within the range of 0.42 to 0.50 psig, with a corresponding output. The theoretical efficiency is between 30 and 40%.

The Output Pressure versus Input Pressure Curve. The curve is based on a supply pressure of 0.46 psig and illustrates the digital characteristics of this device: a small input pressure around 0.017 psig is capable of reducing the 0.17 psig output pressure to 0.

The extremely high sensitivity of this device is illustrated by the curve. The residual noise is negligible. The amplification factor is 10 when the input pressure is 0.017 psig and the output pressure is 0.17 psig.

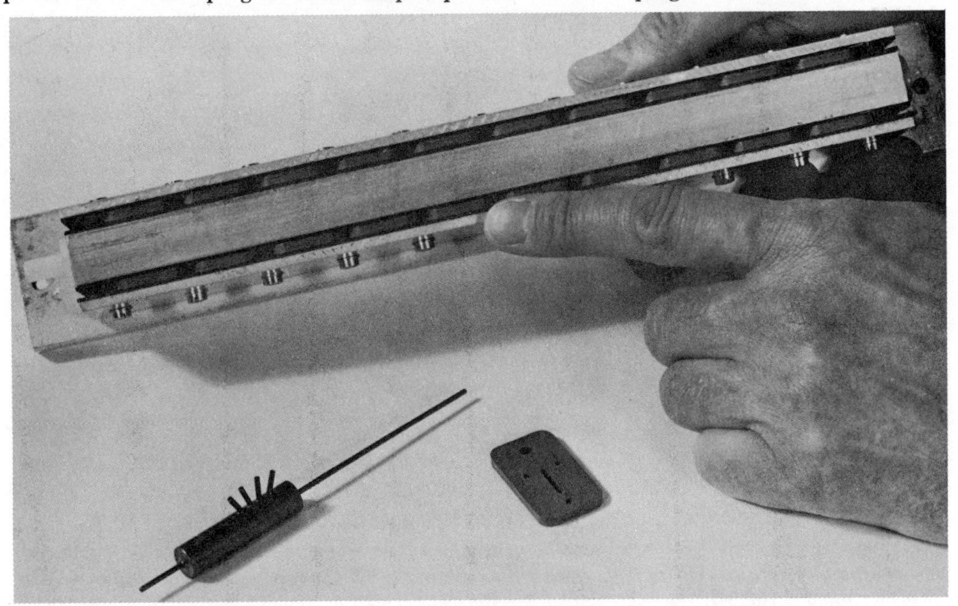

Comparison of a tubular and a planar laminar/turbulent logic device. (Courtesy Maxam Fluid Power, Great Britain.) Toward the topside is a sandwich of planar elements while just below them on the left is a single tubular device alongside its planar equivalent.

3-3-2-3 The Design and Application of Laminar/Turbulent Flow Devices

With only one input the device functions as a NOT (output = 0 if input = 1). It is quite easy to design devices with several input ports. Each input is capable of stopping the output flow by making the laminar stream turbulent.

Figure 3-5 illustrates the two basic laminar/turbulent flow devices, the tubular and the planar. Both types provide the four input NOR logic function: $U = \overline{a + b + c + d}$. The output = 1 only when all four inputs equal 0.

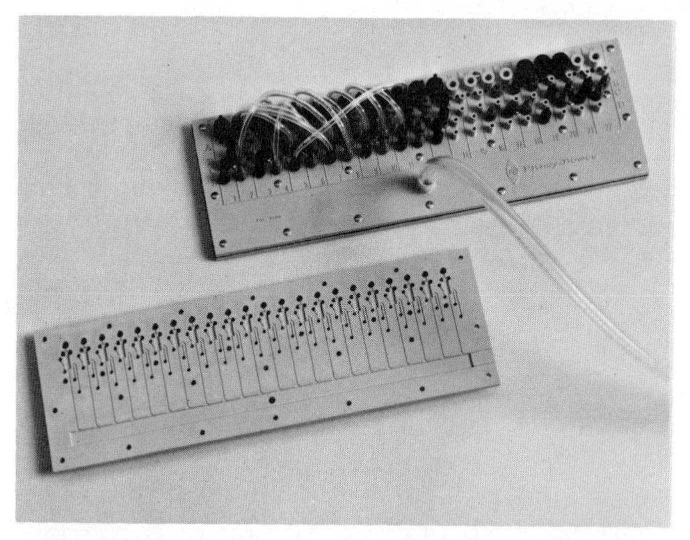

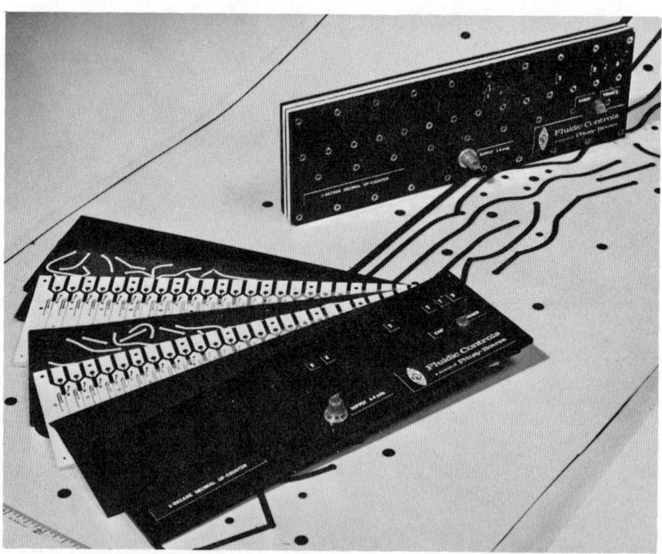

Flowboards and flowbriks. (Courtesy ASCO Fluidics, U.S.A. formerly Pitney Bowes.) The "flowboard" above is an assembly of 22 NOR logic gates which operate in the pressure region of jet destruction devices illustrated in Fig. 3-2. The open half section shows long emitter channels leading to interaction regions. Opposite each emitter channel is a collector channel at the top of each element. Four control channels enter each interaction region from the sides. A common plenum feeds all emitters from the bottom of the flowboard. The "flowbrik" below is a sandwich containing two layers of NOR logic gates interconnected by the gaskets with circuit passages within them. The entire assembly is riveted together to form a solid flowbrik.

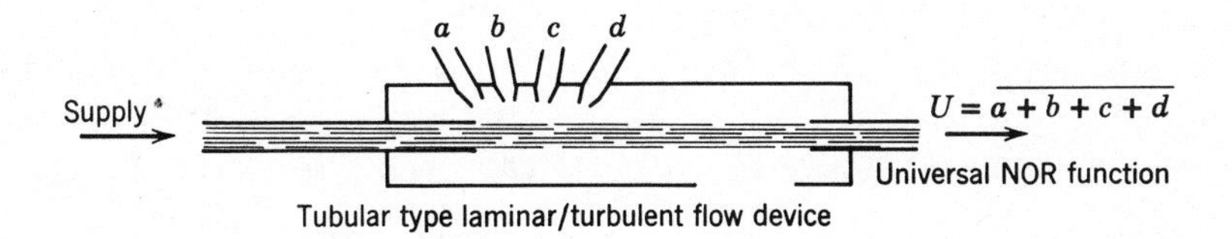

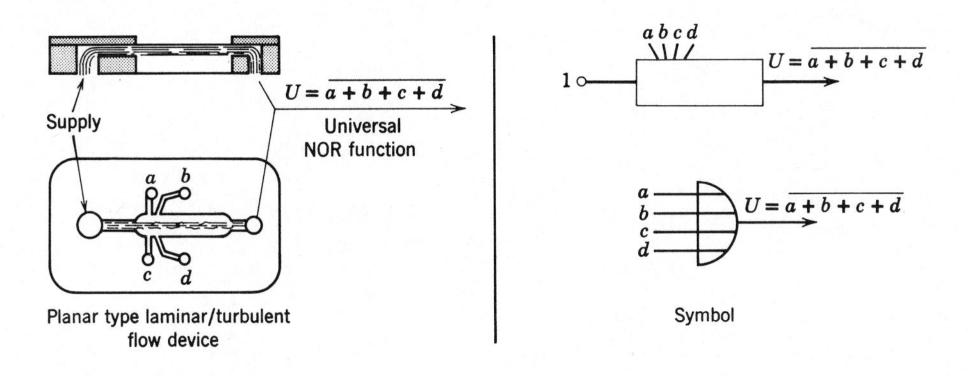

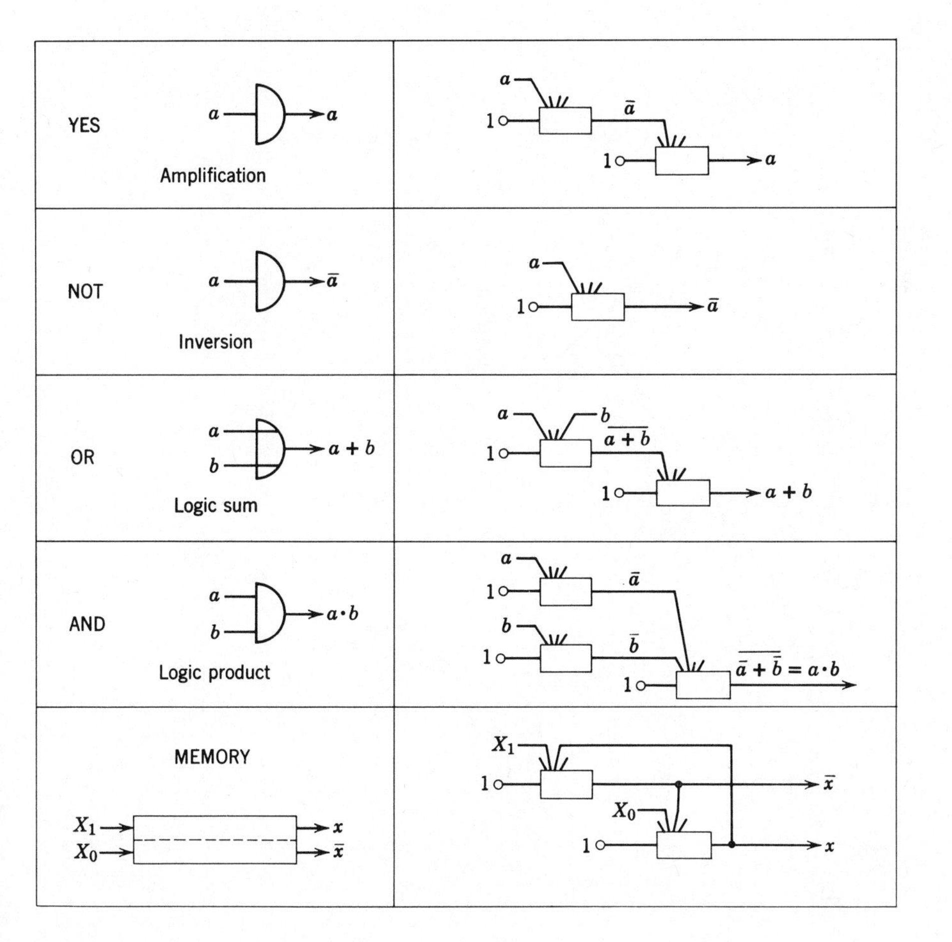

FIG. 3–5. Laminar/turbulent flow devices and their logic functions.

In section 2-4 we described the "universal" type logic functions and their applications. Here, with the symbol usually used to identify laminar/turbulent flow devices, Fig. 3-5 illustrates the creation of basic logic functions with only the NOR universal type device.

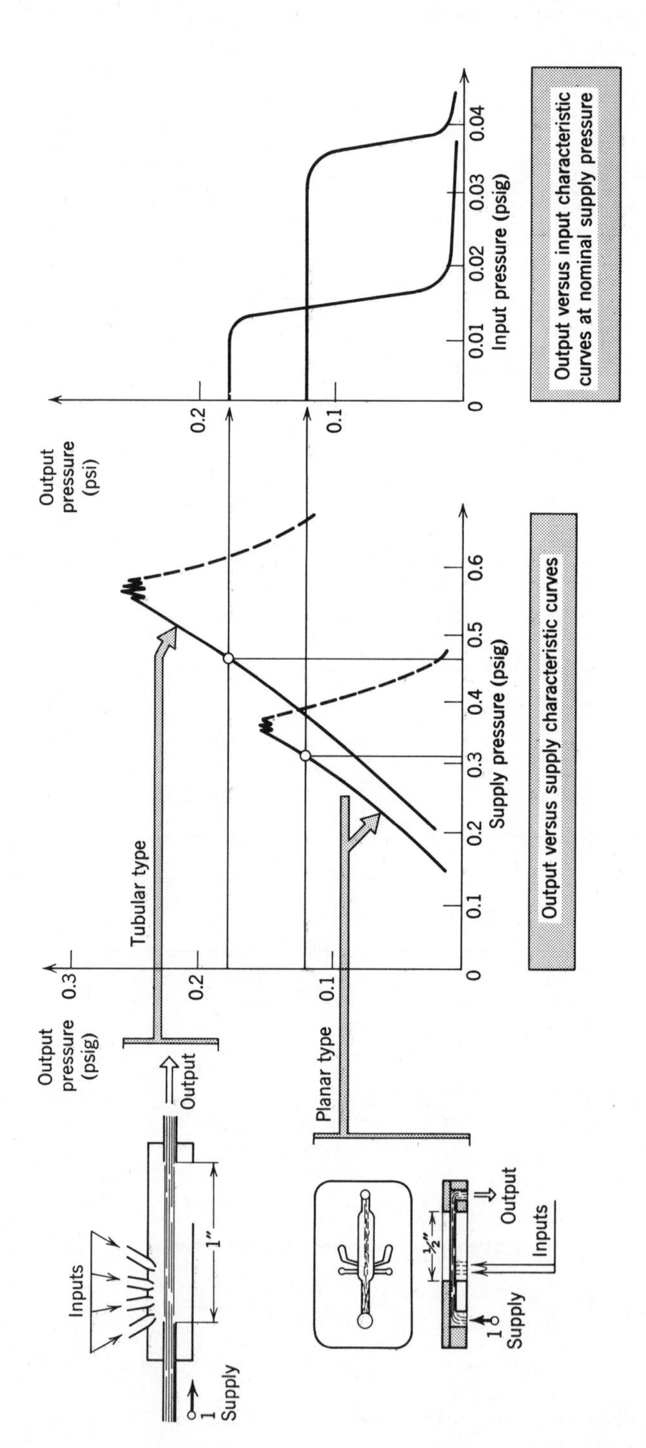

FIG. 3–6. Comparison of two types of laminar/turbulent flow devices: tubular type (large) and planar type (small).

	Tubular type	Planar type
Length of laminar flow	1 in.	1/2 in.
Nominal supply pressure	0.46 psig	0.31 psig
Corresponding output pressure	0.17 psig	0.11 psig
Minimum input pressure	0.017 psig	0.035 psig
Fan-out capability	6 to 10	4
Response time	5 ms	2 ms

The planar type can be stacked in compact configurations that allow either external interconnections or engraved integrated circuits.

Figure 3-6 illustrates a comparison between tubular and planar type devices, with characteristics typical of their kind. The miniaturization of planar-type devices is more easily implemented than for tubular types; however, the miniaturization process places restrictions on the operating characteristics:

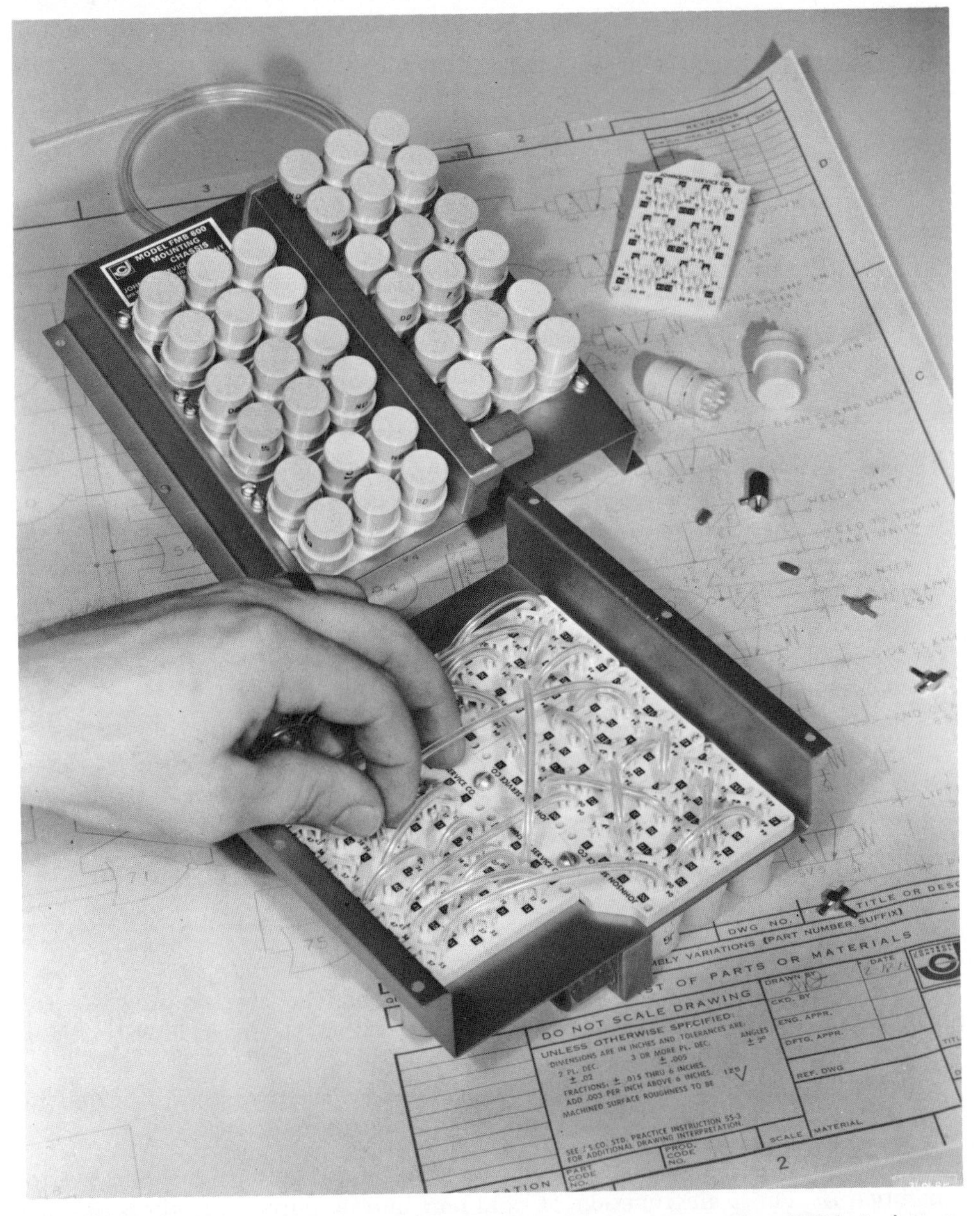

Impact modulator logic system (Courtesy Johnson Service Company, USA.) These NOR logic elements are classified as "jet destruction" devices in Fig. 3-2. They operate according to the basic principles illustrated in Fig. 3-8. This view shows devices assembled onto plug-in cards which attach to panels with circuit connections on the backside.

- The operating pressure must be reduced when the flow passages are reduced in size.
- Response times are improved but sensitivity and fan-out capabilities are decreased.

The laminar flow characteristic of these devices becomes unstable when the device is subjected to mechanical or audible vibrations around 5,000 Hz. Vibrations within this region are common in industrial environments. However, it is relatively easy to mount the devices within sealed industrial enclosures that provide isolation from this potential hazard.

The power consumption of these devices is very low, less than 1 W per device. This is a direct result of the small diameter of the flow passages which in turn require a low pressure supply to maintain the laminar flow characteristics.

In Chapter 4 we describe the peripheral equipment required in constructing complete systems.

The output of the logic must be amplified to standard pneumatic pressure levels if it is to be used to actuate power devices. On the input side of the devices the inherent high sensitivity of laminar/turbulent elements is one of the major features that make them desirable as an industrial logic device.

3-3-2-4 *The Operating Principle of Impact Modulators*

Figure 3-7 illustrates the operating principles of the active type device. In contrast to laminar/turbulent flow devices, impact modulators do not employ laminar flow phenomena; rather all flows are turbulent.

Two supply jets are required to create an output signal: Jets A and B directly oppose each other to form an impact plane. Supply jet B is somewhat weakened so that the impact plane will be formed off center within chamber B. The supply flow is directed to the output port within chamber B when no control signals are "ON."

When a control signal is applied, it is directed against supply jet A, thereby weakening it, so that the plane can no longer be formed within chamber B. Supply jet B now exits through the orifice between A and B and vents to atmosphere. The output now switches "OFF."

3-3-2-5 *Construction and Application of Impact Modulators*

Figure 3-8 illustrates the design of a typical impact modulator. The device consists of the following:

- Only one external supply port is required, the two opposing jets are created by the internal configuration of flow passages which includes a restricor to weaken supply jet B.
- The output of the device is controlled by an input control signal to one or more of the four individual inputs (a, b, c, or d).

The device can have several common output ports, all connected to chamber B. The logic function of common output ports U is the same as for the laminar/turbulent flow devices: the universal NOR logic function $U = \overline{a + b + c + d}$.

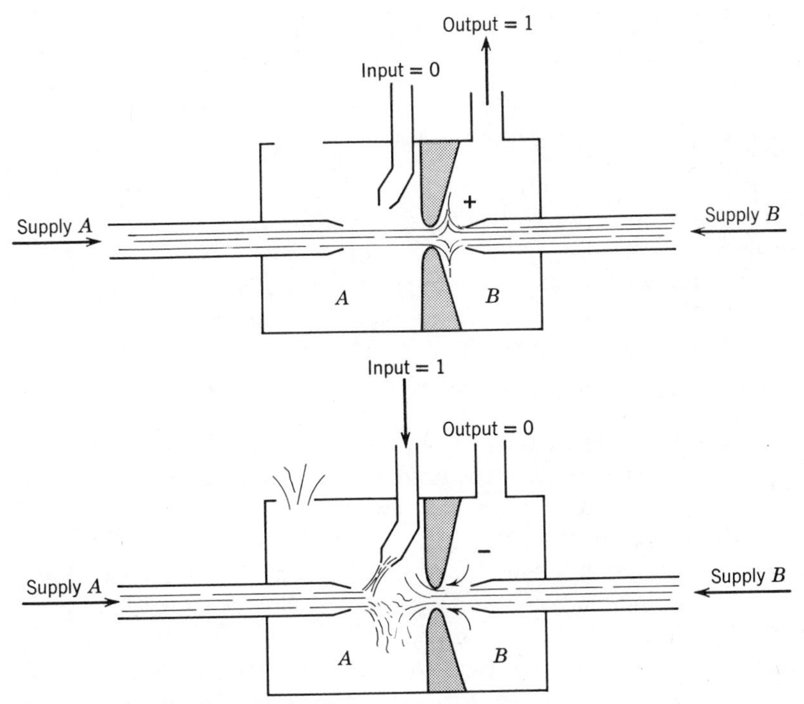

FIG. 3-7. The operating principle of impact modulators.

Figure 3-8 illustrates the method of creating the basic logic function with this NOR logic device. The method is similar to the one illustrated on Fig. 3-5 for the laminar/turbulent flow devices.

Since impact modulators do not require a laminar flow as do laminar/turbulent flow devices, their operating pressure range is less limited. The device illustrated operates on a nominal supply pressure of 1 psig. The usual supply pressure range is between 0.5 to 5 psig.

The switching characteristics of these devices are good, but do not equal the digital characteristics of typical laminar/turbulent flow devices. The difference is due to operating principles which result in slightly slower response times. Also the amplification factor is not so great as laminar/turbulent flow devices. However, the efficiency and fan-out are usually higher.

The following table illustrates typical operating characteristics.

Supply pressure (psig)	Number of devices controlled by the concerned device	Output pressure (psig)	Minimum input pressure (psig)
1	0	0.46	0.014
	5	0.34	0.010
	10	0.20	0.010

Air consumption is relatively low (0.2 W at 1 psig supply) and a nominal response time of 3 msec.

3-3-3 Jet Deflection Logic Devices

This type of device uses input signals to divert the output back and forth between the two output ports. The jet is not destroyed as in the jet destruction type of device.

According to the means employed, jet deflection devices are divided into the following types: jet interaction, wall attachment, (the most popular) vortex feedback, and wall reflection.

3-3-3-1 History

A large amount of research has been concentrated on this group of devices, much of this has been government funded in many of the industrial countries.

The original research on wall-attachment devices came from Henry Coanda, a Roumanian-born aerodynamics engineer, who moved to France in the 1930s. There he published his findings on "the wall-attachment phenomena." This principle has become known as the "Coanda effect."

R. E. Bowles, an American engineer, was the first to apply the "wall-attachment principle" to logic devices in the late 1950s. United States manufacturers have since developed many logic systems employing jet deflection devices. Examples of these are Corning Fluidics, a division of Corning Glass which employs a glass photo engraving; General Electric Fluidic division which employs metal etching, and Bowles Fluidics which employs plastic injection molding. Similar notable developments in Europe were achieved in Italy by RIV-SKF, in England by Plessey, and in France by Bertin.

3-3-3-2 Operating Principles of Jet Deflection Devices

Figure 3-9 illustrates the principles employed to deflect a jet.

Jet Deflection with Jet Interaction Effect. A simple method of jet deflection is to use a side positioned input control jet. This process can also be used for analog controls. The degree of deflection is directly proportional to the power of the input control jet. Later we discuss how this principle is applied to passive logic devices.

Jet Deflection with Vortex Feedback Effect. Figure 3-9 illustrates how jet deflection is created by means of an auxiliary jet taken from the main jet and redirected to interact with the main jet by creating a vortex.

Jet Deflection with the Wall-Attachment Effect. Figure 3-9 describes wall attachment or Coanda effect. The free jet entrains atmospheric air from within the passageway thereby creating a depressed area (a partial vacuum). The differential pressure on both sides of the jet causes it to bend toward the depressed area and attaching itself to the wall.

Figure 3-10 illustrates the application of these principles to the various types of active jet deflection logic devices.

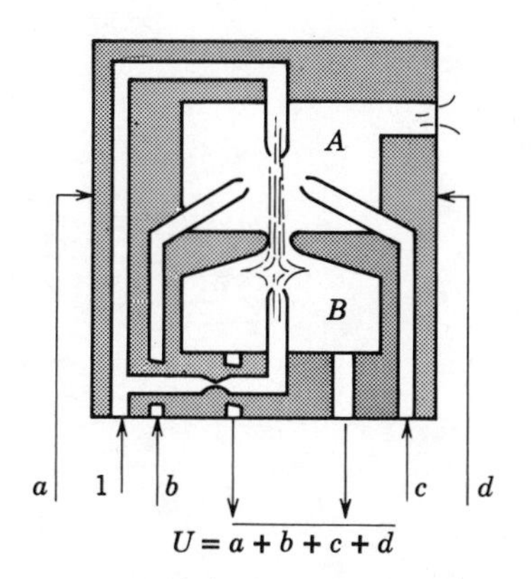

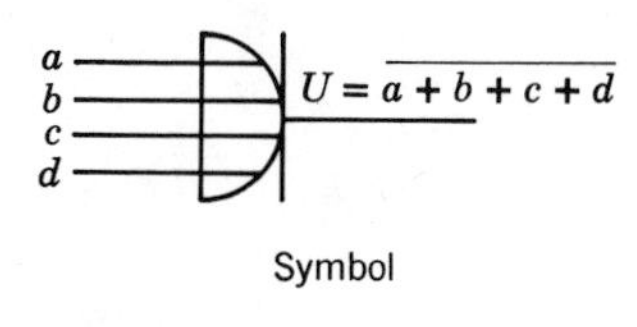

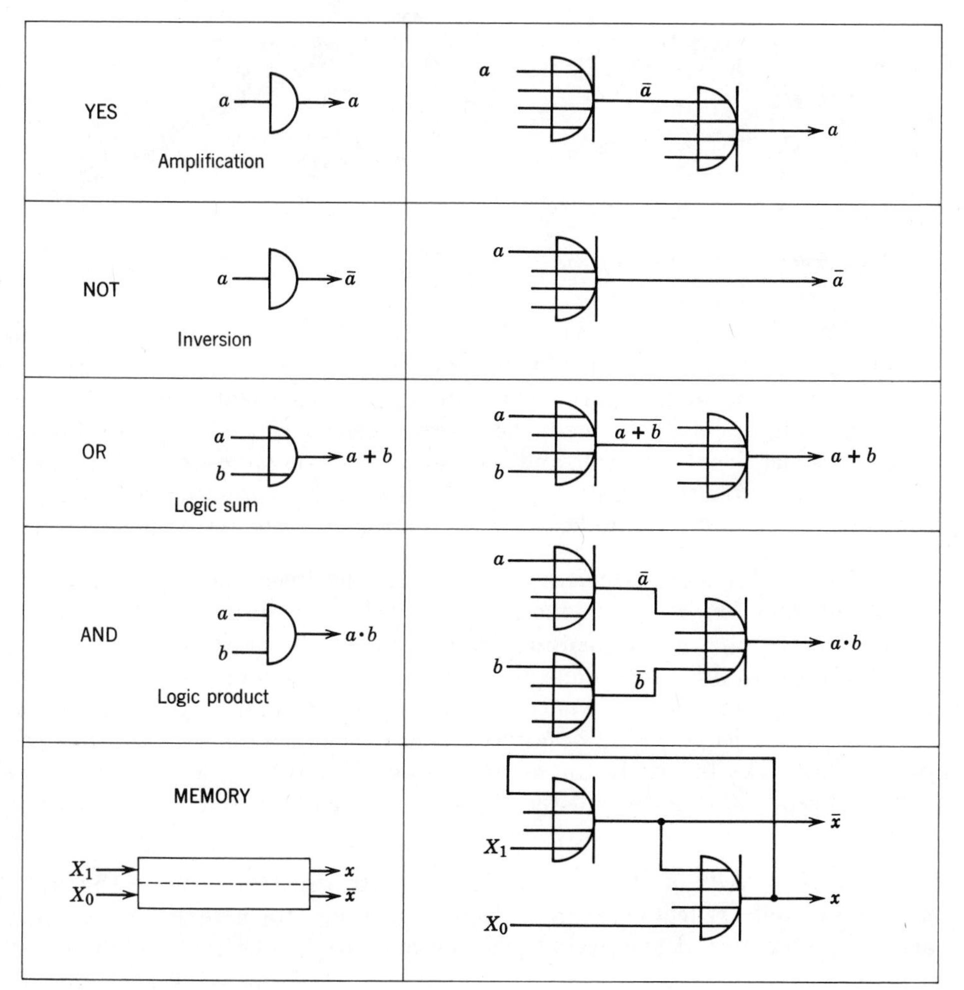

FIG. 3–8. Impact modulators and their logic functions.

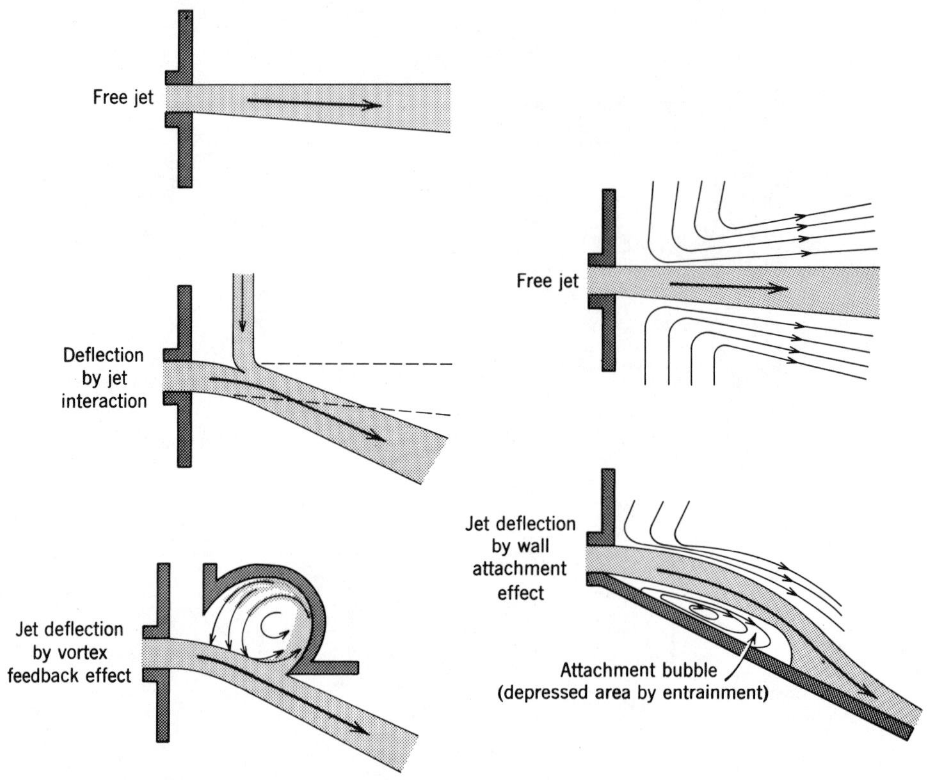

FIG. 3–9. Principles used to deflect a jet.

Wall-Attachment Devices. In the absence of an input signal the supply jet attaches itself to one of the walls, thereby channeling the supply to the corresponding output port. When an input signal is applied to the side where the supply jet is attached, the depressed area is filled, thereby disconnecting the jet from that wall and switching it to the opposite wall where it adheres, even after the input signal is removed. The depressed area is established on the opposite wall to hold the jet and the opposite output is now "on."

Vortex Feedback Devices. The vortex feedback device creates the equivalent function by employing an entirely different principle. The central interaction chamber is designed to create a vortex that bends the main jet as a result of positive pressure on one side of the jet and a vacuum on the opposite side of the jet created by a venturi effect. When an input signal appears on the side of the jet with the negative pressure, the vacuum is filled and the pressure switches the jet to the opposite side of the device where the vortex principle establishes itself and the supply jet exists from the opposite output port.

Wall-Reflection Devices. Figure 3-10 illustrates how the four principles—wall reflection, vortex feedback, wall attachment, and impact action—are combined and simultaneously employed to switch the supply jet

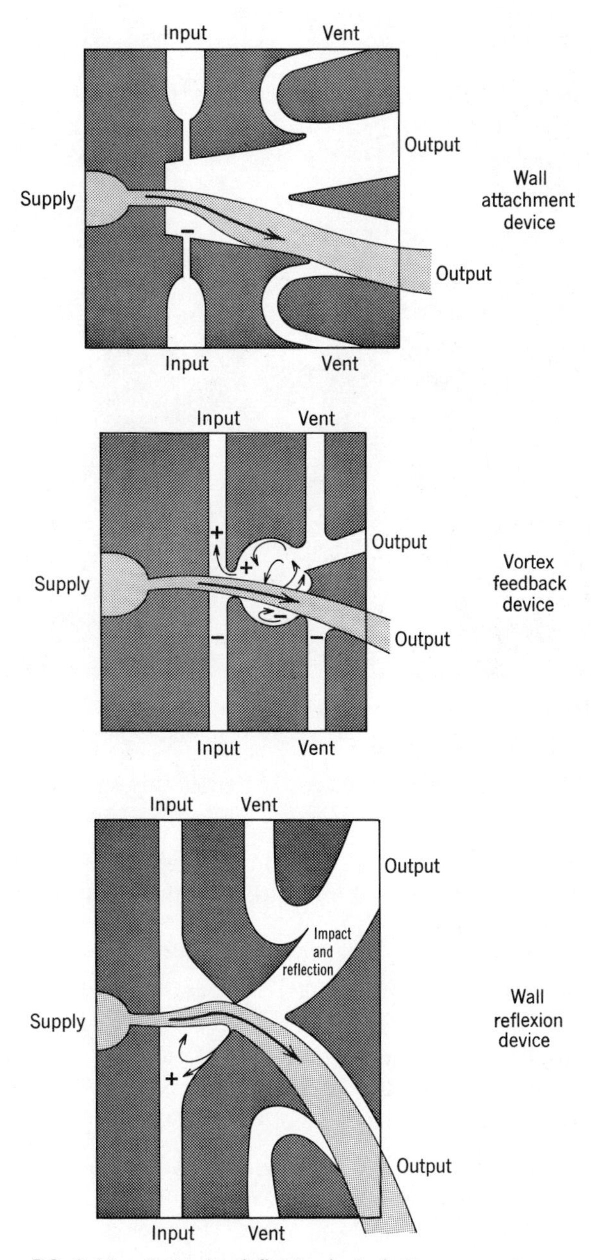

FIG. 3–10. Active jet deflection logic devices.

when an input signal is applied. If the jet could be seen in action, it would appear as a flexible blade that switches back and forth between its two equilibrium positions in response to the input signals.

The wall reflection device is the only one that switches the supply jet to the side of the device that receives an input signal. All other devices discussed, switch the supply jet to the side opposite the input signal.

Natures wall attachment. (Courtesy Corning Fluidics, USA, Sovcor, France.) Spilled water clings to the chin of this child serving to visualize wall attached flow in its natural form.

Figure 3-11 illustrates the operating principles of jet deflection passive logic devices.

- The OR device (inclusive) employs a very simple principle: an input signal from port *a* or *b* or both is guided through the device to a single output port.
- The AND device requires input signals *a* AND *b* simultaneously, and of equivalent levels. The interaction of the two jets creates a blending action resulting in a single jet which is guided through a central channel to the output port. If either of the input signals (*a* or *b*) is applied alone, the input will flow directly to its corresponding vent, which exhausts the signal to atmosphere.

The AND–EXCLUSIVE OR device has two outputs U_1 and U_2. If input signals *a* OR *b* is applied alone, an output will result at port U_1 only, thereby creating an exclusive OR function. If input signals *a* AND *b* are applied simultaneously an output will result at port U_2 only, thereby creating the AND function.

3-3-3-3 *Design and Application of Jet Deflection Devices*

In theory it is possible to apply all the devices above in combination to create specific circuits; however, before this can be accomplished in practice, two problems must be overcome.

Output Loading Effects. The operating media employed in a device to create an output can be allowed to flow freely through the device or be restricted, depending on the needs of a circuit at a particular time.

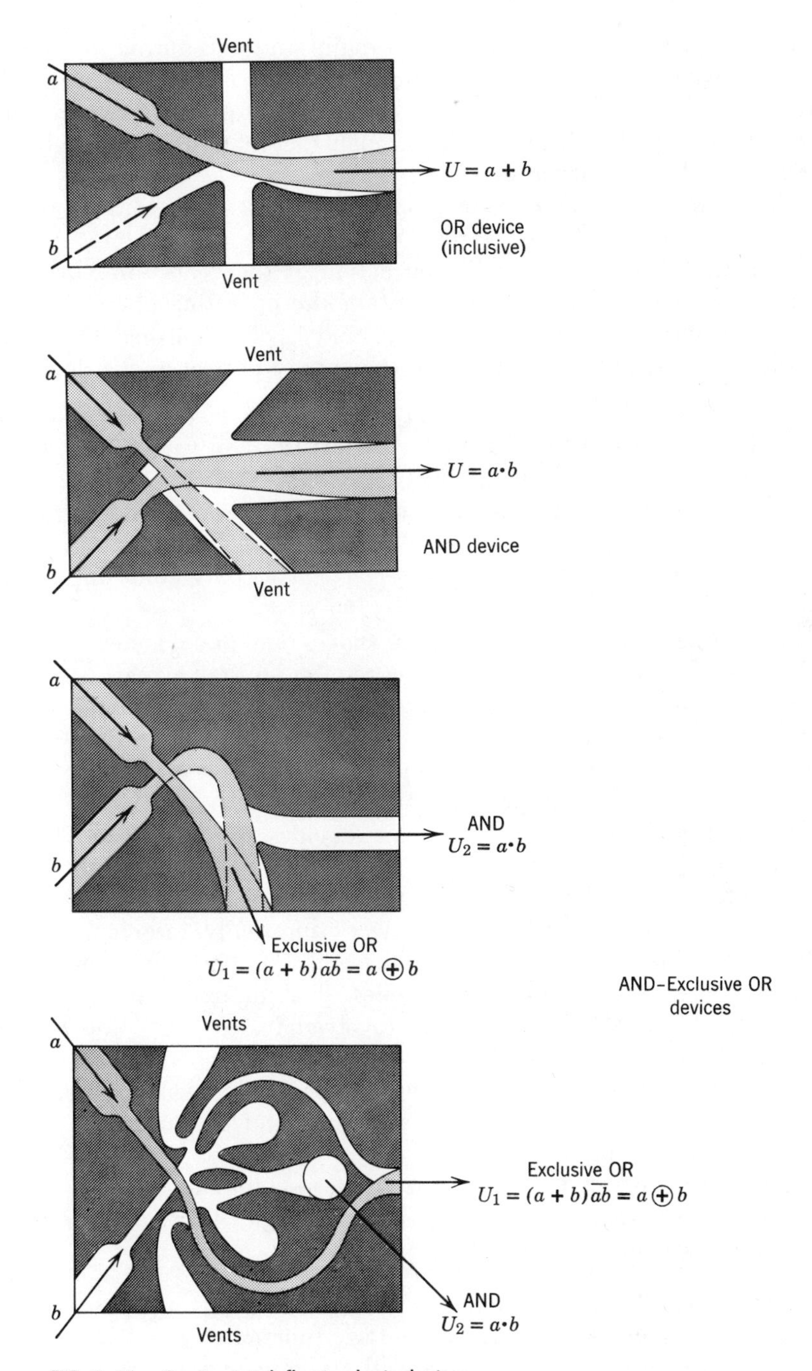

FIG. 3–11. Passive jet deflection logic devices.

The operating phenomena of jet deflection devices are based on the necessity of continuous flow through the devices. This requirement cannot be altered if proper operation is to continue; even if the device receiving, the output (the load) signal cannot consume the flow, for example, a diaphragm relay after actuation.

The purpose of the vent ports is to maintain the required internal pressure balance when the load stops consuming the flow. The vents are designed as branches off the output passages so that excessive pressure rise in the passages will be diverted to atmosphere through the vents, thereby maintaining the correct internal balance of pressure.

There are two types of vents, both have descriptive names based on their shapes—"diode" vents or "vortex" vents.

Even with the pressure relieving action of the vents, pressure adjustments are necessary sometimes to maintain the operating phenomena. This type of adjustment is known as "tuning" and is accomplished by means of adjustable flow restrictors connected to the supply port of active devices.

Input Signal Levels. Active jet deflection devices are usually good input signal amplifiers. Their fanout characteristics can vary between factors of 3 to 8 depending on the device design.

Reliable operation of these devices is possible over a wide input signal level range.

In contrast to active devices, passive devices only guide input signals through to the outputs; they do not amplify.

On another hand, we have seen that some passive devices require equivalent input signal levels to create the required output signals.

A well-designed logic control system is the result of compatible combinations of active and passive devices, the active devices amplifying and the passive devices consuming the logic signals. The proper use of flow restrictors assists in keeping the internal pressure balances.

When using nonmovable-part logic devides a choice must be made between the methods of system design: individual logic functions as we have discussed in this chapter, or special integrated circuits. The integrated circuits consist of layers of individual logic devices internally interconnected and molded or assembled into neat, compact assemblies. These custom-made circuits are not advantageous unless multiples of 20 or more are required, because of the additional costs and time involved in designing and manufacturing them.

When one of a kind (or only a few) of a particular system is required, it is more economical, faster, and easier to design and to build with standard individual logic functions.

Standard integrated circuit assemblies—counters, timers, and sequencers—can be used to supplement the basic logic function devices to reduce the overall size of a control system.

Figure 3-12 illustrates a logic system employing two types of wall-attachment devices to create all of the basic logic functions.

The bistable device is the active wall attachment element described in Fig. 3-10. It provides the means to create the memory function: the supply flow can be directed to either of the two outputs x or $\bar{x}$ by input signal pulses to X_1 or X_0. The output is maintained at the last port selected even after the input signal is removed.

The monostable device is a slightly asymetrical version of the bistable device. The device is designed so that in the absence of both input signals the output $\overline{U}$ will always be "ON." Input signals are channelled through an integrated passive OR. When inputs a OR b (or both) are receiving signals, the output of the passive OR switches the flow to output port U.

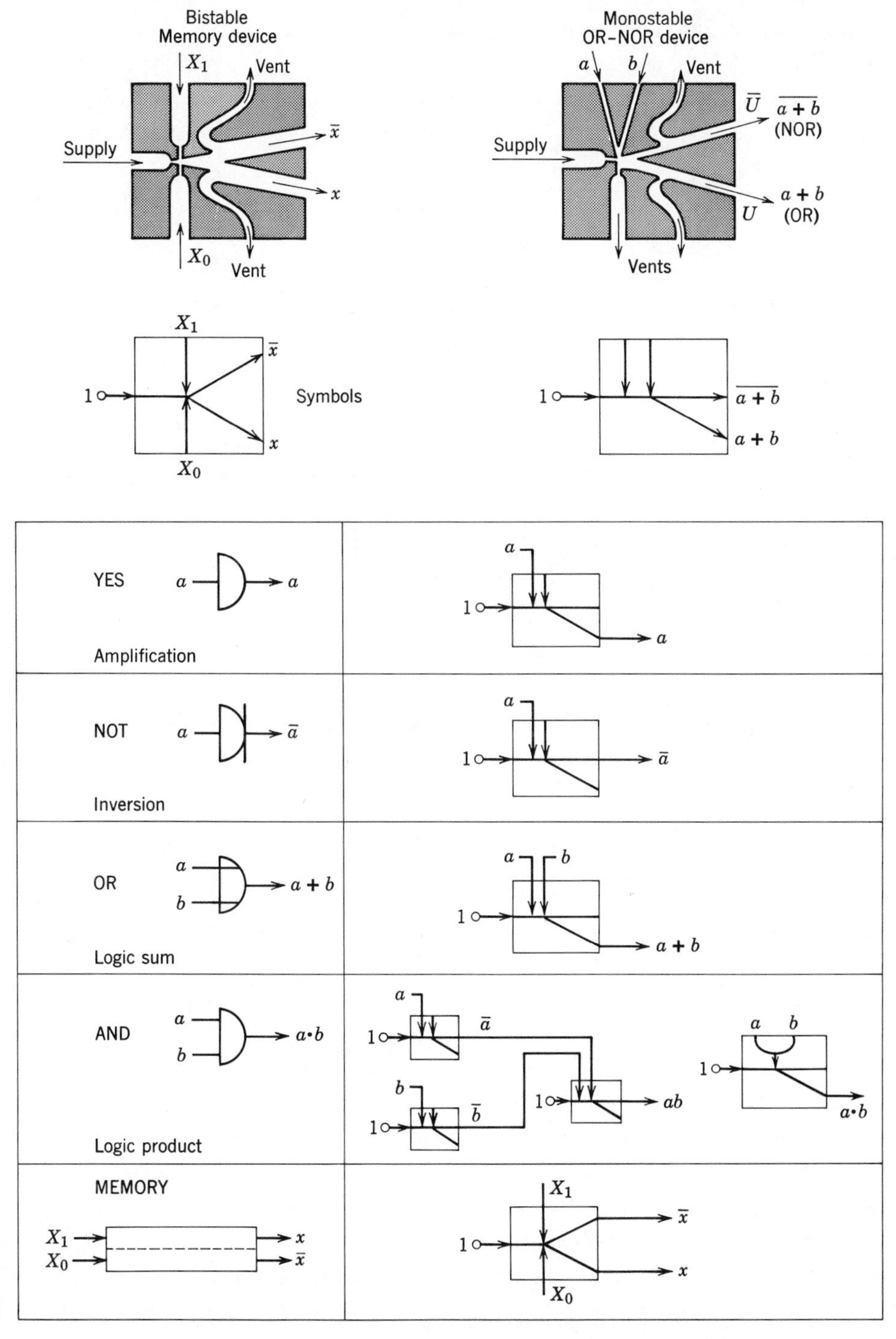

FIG. 3–12. Bistable and monostable wall attachment devices, and their application.

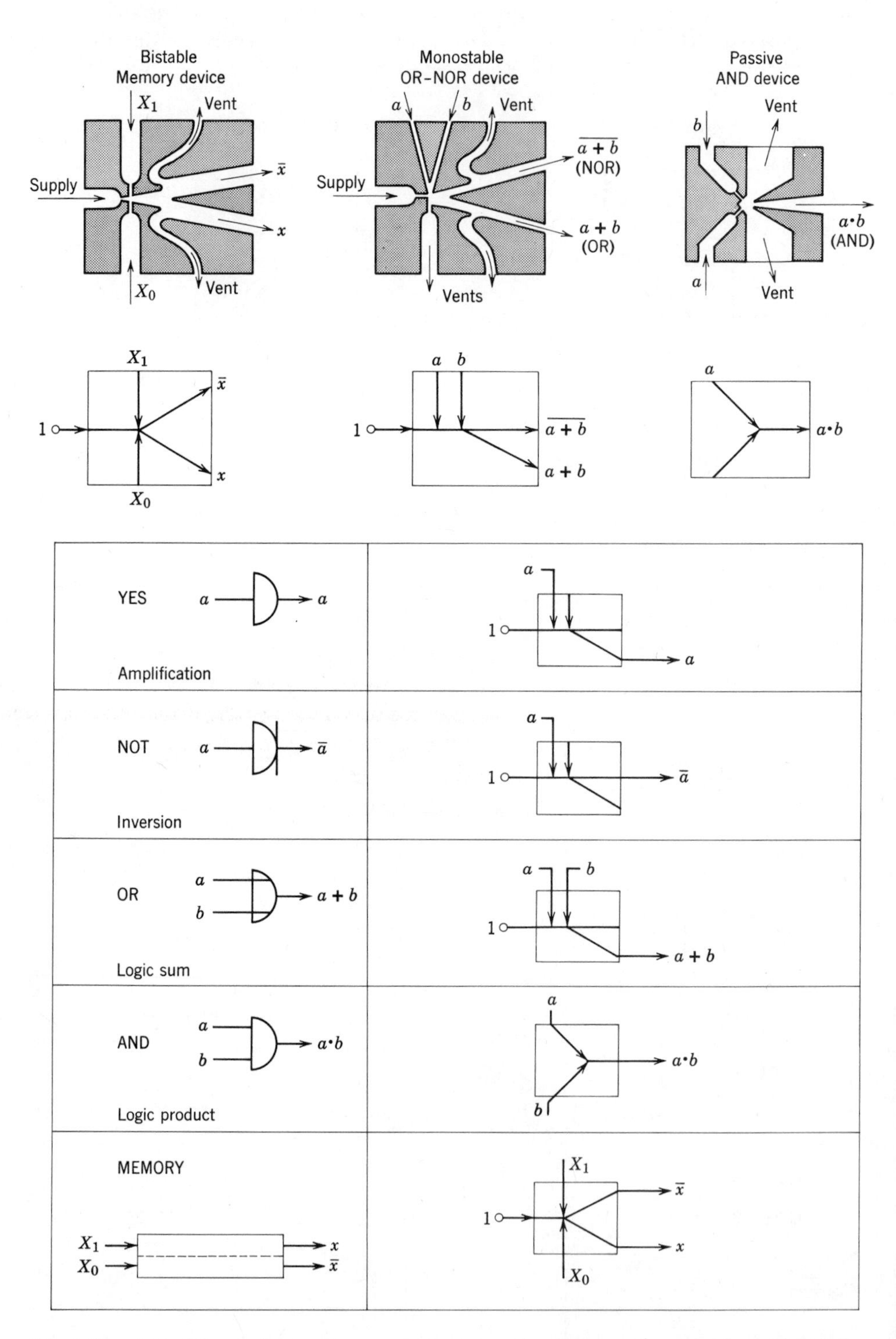

FIG. 3–13. A jet deflection logic system employing two active devices and one passive jet interaction AND device.

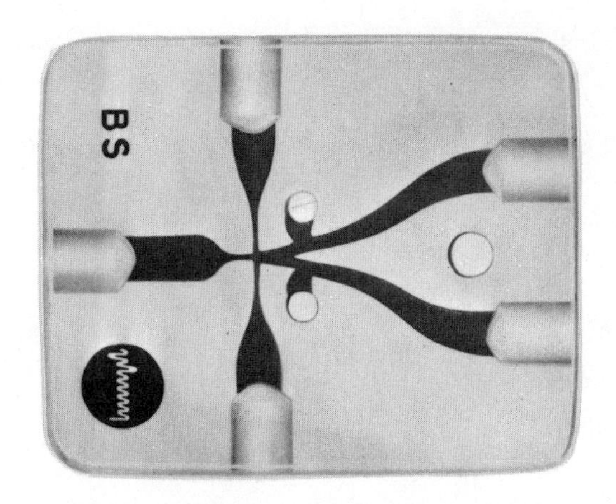

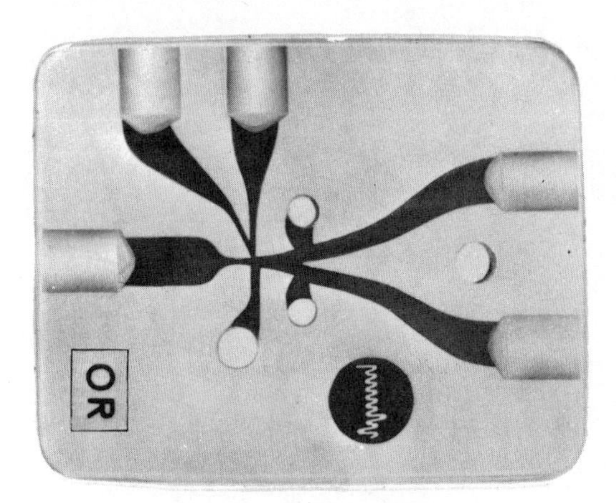

Basic devices in a jet deflection system. (Courtesy Plessey, Great Britain.) The devices are molded from transparent plastic which leaves the internal shapes quite visible. From the top, they are an active bistable wall attachment memory element, an active monostable wall attachment OR-NOR gate, and a passive jet interaction AND gate. This system is interconnected with plastic tubing that plugs into sideports. Operation is according to the principles described in Fig. 3-13.

Baseplate mounted wall attachment system. (Courtesy Corning Fluidics, USA.) The basic operating principles of some of these devices are illustrated in Fig. 3-12. The etched glass elements are fastened, O-ring sealed to the baseplate where circuit connections are made to the backside.

This active monostable device is also known as an OR–NOR device because of the following:

- Output port U is "ON" when inputs a OR b are receiving signals.
- Output $\overline{U}$ is "ON" when neither input a NOR b are receiving signals.

Figure 3-12 illustrates the versatility of this multifunctional device. It can create the YES, NOT, OR, NOR functions with a single device and the AND function with three duplicates of itself. To eliminate the need for three devices to create the AND function, manufacturers of these particular devices have developed a single device to perform this function.

A simpler solution for the system above is to employ a passive jet interaction AND device. This system is illustrated in Fig. 3-13. We have seen that passive AND devices required the use of inputs signals of equivalent levels. Circuits with this system will require pressure compensation to obtain proper operation.

Other complete systems could be described; however, they all operate similarly to wall-attachment devices, the only differences being the characteristics of the basic switching phenomena; some employ vortex feedback, others use wall reflections.

Operating Characteristics and Performances. The primary characteristic of jet deflection active logic devices is the supply passage cross-sectional area. This area and the supply pressure level determine the power consumption and the output power level.

Figure 3-14 compares the performance data of active monostable devices employed in systems currently available.

Active device Type	Supply passage cross-sectional area	Supply pressure (psig) –Maximum –Nominal –Minimum	Minimum control pressure (psig) at nominal supply pressure	Output pressure dead-ended (psig) at nominal supply pressure	Amplification factor $= \frac{\text{output } P}{\text{input } P}$	Fanout factor	Response time (ms)	Consumption with nominal supply pressure
1	0.010 × 0.020 nominal	10 5 3	0.5	1	2	3-4	2.5	0.06 SCFM
2	0.100 × 0.010 nominal	15 10 7.5	1.2	2	1.8	2	5.0	0.04 SCFM
3	0.014 × 0.017 nominal	1.5 1.25 1.00	0.2	0.5	2.5	2	1.0	0.06 SCFM

FIG. 3–14. Operating characteristics of typical wall attachment devices.

The Manufacturing of Devices. To create and to maintain the operating phenomena and performance characteristics of nonmovable-part devices, the internal chamber and channel configurations must be held within strict tolerances. The following manufacturing methods have been employed to date:

- Photoetching of glass permits the manufacture of devices with high dimensional stability and the ability to hold close tolerances. Also the glass is immune to most industrial environments and is suitable for small quantity production. Glass is the best material when complex or multiple circuits are required, even though it is more costly to process.
- Plastic injection molding is the most economical process for quantity productions.
- Powder metallurgy has also been used occasionally.

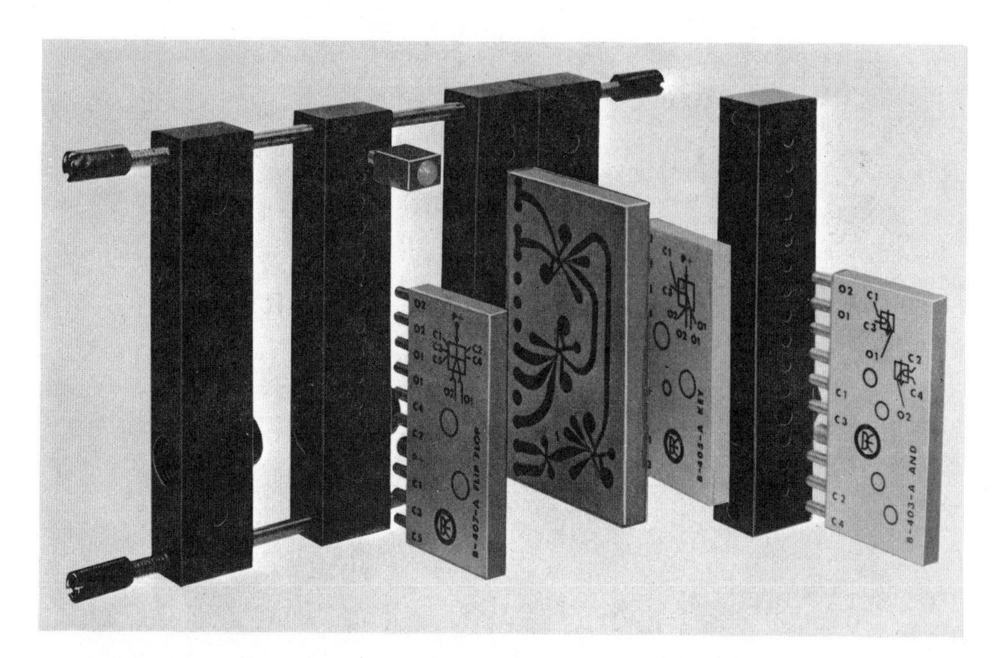

Plug-in modules of a jet deflection system. (Courtesy Bowles Fluidics USA.) **Baseplates with sockets to accept the plug-in modules are hangered on rods to form panel assemblies. The baseplates plug into each other thus transmitting supply power to each other. Circuit connections are made to the baseplates from behind the panel with plastic tubing.**

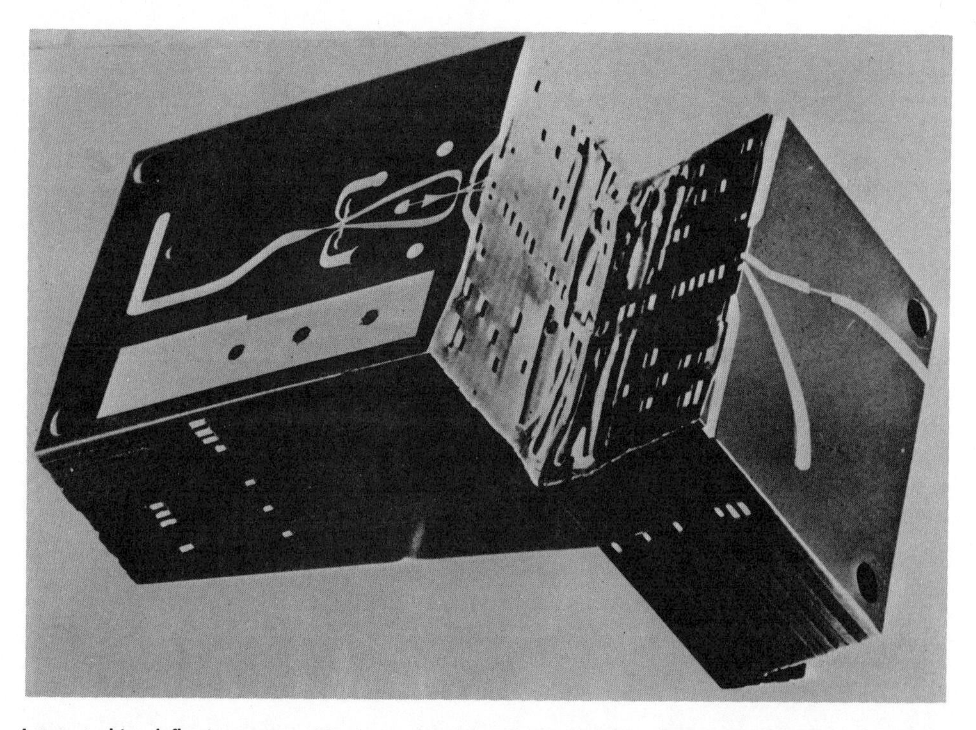

Integrated jet deflection system. (Courtesy of Corning Fluidics, USA.) **This cutaway section shows how layers of glass with etched circuit connecting channels fuse devices into a single integral monolithic block.**

The Packaging of Devices and Interconnections. The photographs illustrate some of the basic methods of device construction. If individual components are employed, they can be interconnected individually or subbase mounted. When complex circuits are required, completely integrated-circuits can be employed, which can be constructed in single or multiple layers or blocks.

3-4 MOVABLE-PART LOGIC DEVICES

3-4-1 History

The manufacturers of moving parts logic devices are divided into two basic divisions: high-pressure and low-pressure devices

Low Pressure Movable-Part Devices. Originally these were designed as logic complements to pneumatic analog instrumentation systems, operating between 1 and 15 psig, and occasionally up to 30 psig.

Current low-pressure movable-part devices operate in the 1 to 30-psig range. Their small flow capacities result from short strokes of the poppets or diaphragms. They usually employ techniques developed for analog pneumatic instrumentation controls. Their regions of origin include the following countries:

- The USSR where many industries use a three-diaphragm device developed at the Institute of Control Sciences in Moscow. Known as USEPPA in the USSR, this system is licensed for manufacture in Italy where it is called SELP.
- East Germany (German Democratic Republic) where VEB Reglerwerk manufactures a system based on a double diaphragm logic element for their system called DRELOBA.
- Czechoslovakia where the Industrial Automation Works manufactures a system called Pneulog originated by the Technical University of Prague.
- Hungary where MMG manufactures a three-diaphragm device developed at The Automation Research Institute in Budapest for their system called TRIMELOG.
- West Germany (Federal Republic of Germany) where two companies Festo and Samson manufacture systems employing multiple diaphragm devices.
- Great Britain where two companies Techne and Air Automation produce logic elements and systems.
- The United States where the company Double A manufactures a multiple diaphragm logic module developed from a concept originated by IBM.

High-Pressure Movable-Part Devices. Originally standard pneumatic power valves, designed to actuate cylinders and air motors, were employed to create logic functions. They operated on standard pneumatic pressures of 30 to 125 psig. Some applications required pressures up to 200 psig. These included:

- The LECQ Company of France used pneumatic poppet valves as early as 1930 to create complex automation systems for coal mines.

- The Martonair company of England first defined the "cascade method," which utilizes standard four-way pneumatic valves to implement fluid logic controls.

The first companies to manufacture successfully and market miniature pneumatic valves were Clippard and Mead in the United States, Kunke of Western Germany, and Universo of Switzerland. However, the application of these valves as logic devices was not readily accepted at the time of their introduction.

Not until the early 1960's were valves developed and marketed primarily as logic devices. These pioneer companies included: Cpoac (France); Aro (United States); Climax France (France); Pneumaid (England); Jouvenel and Cordier (France); and Herion (West Germany).
These were followed by Double A (United States); Crouzet (France); Lang (England); Festo (West Germany); Numatics (United States); Pneucon (United States); Parker Hannifin (United States); Genicon (England); Sempress (Holland); Fluicon (United States); Dynamco (United States); and Clippard (United States).

3-4-2 Operating Principles of Movable-Part Devices

The switching of the fluid is obtained by a mechanical displacement. There is no continuous flow as in contrast to nonmovable-part devices. The signals are isolated from the environment; thus they can be measured as static pressures inside tubes and devices.

Figure 3-15 illustrates the basic differences between the mechanical switching elements commonly used in pneumatic and electrical control devices.
In electrical switching devices:

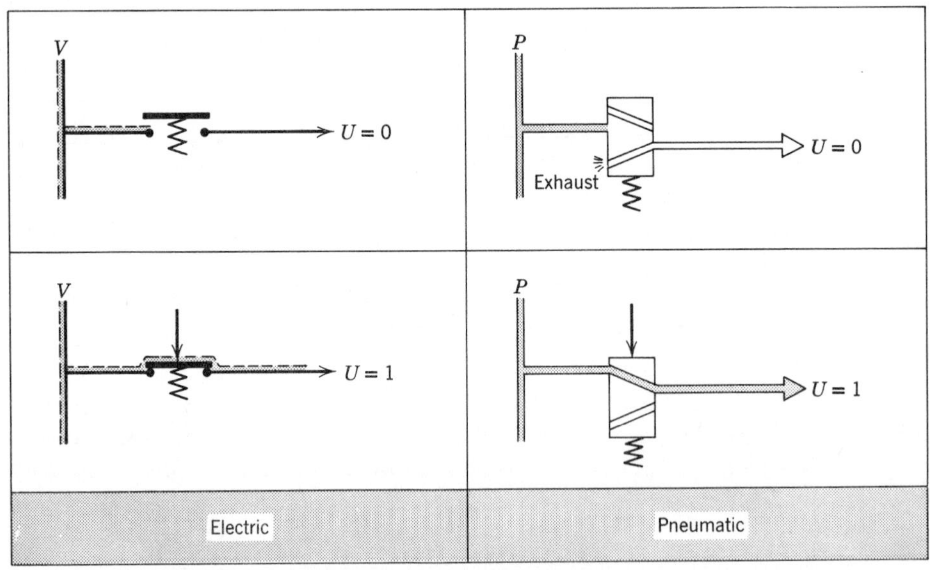

FIG. 3-15. A comparison between electric and pneumatic switching.

- When control contacts are closed, a path is provided to conduct the electrical supply to the output (state 1).
- When contacts are opened, the path is eliminated (state 0).

In fluid switching devices:

- State 1 is obtained by closing the exhaust and opening the supply to the output.
- State 0 is obtained by closing the supply and opening the output to exhaust.

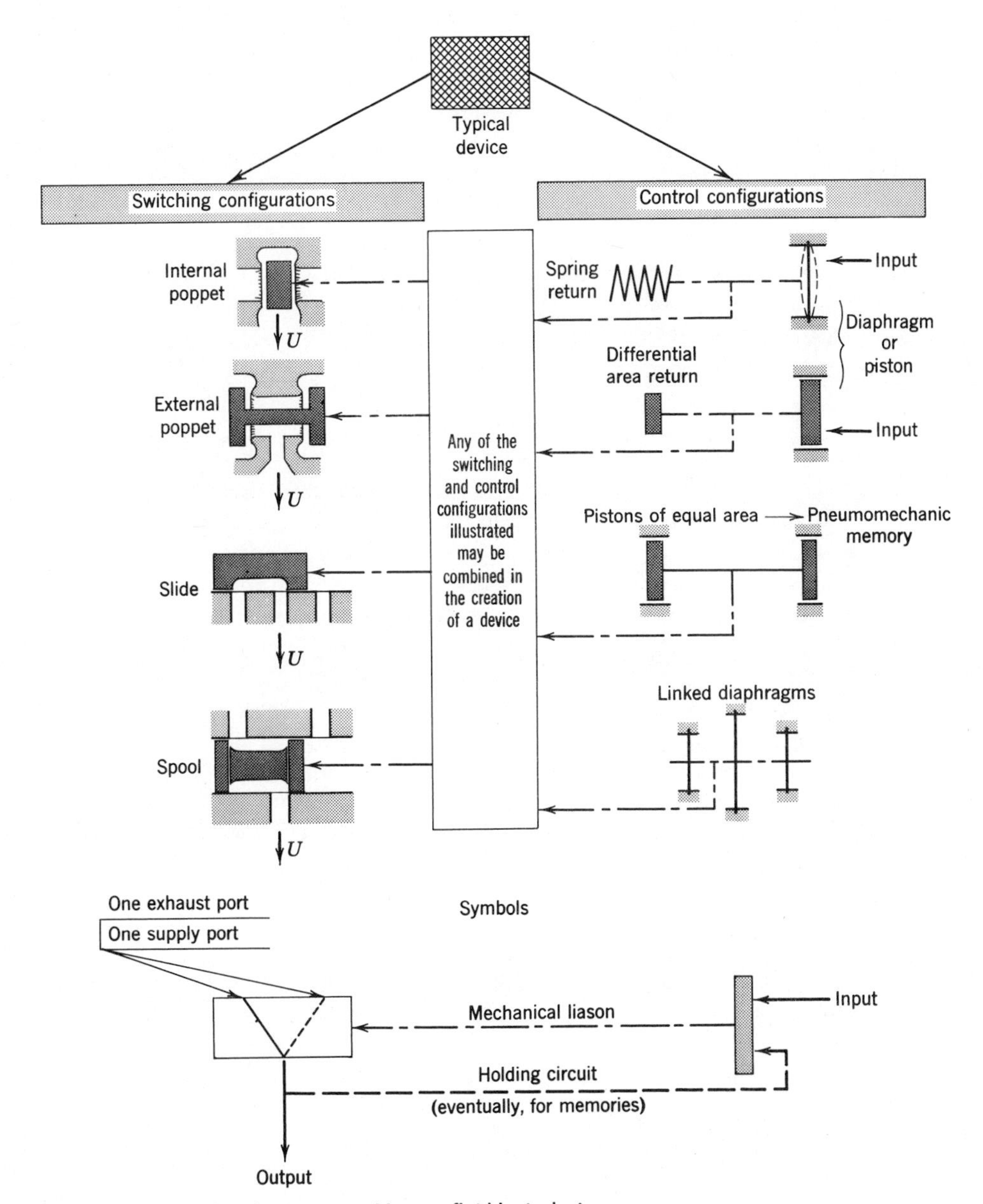

FIG. 3–16. Principles of active movable-part fluid logic devices.

Thus for fluid switching in circuits isolated from the environment, it is necessary to have devices that provide two flow paths. They will be either double poppet devices, or three-way spool valves (that really provide only two flow paths).

Figure 3-16 compares the various principles employed in the design of active movable-part logic devices.

The switching configurations can consist of spool, slide, or poppet. Whichever configuration is used, the center port is always the output. Either of the adjoining ports can be the supply. The third adjoining port always becomes the exhaust.

When the input control signal requirements are satisfied, the switching component is actuated, thereby creating a specific logic function. A large variety of control configurations are available.

Frequently the control and switching elements are integrated into a single assembly. To facilitate comprehension we always differentiate between the control and switching elements in each device.

Figure 3-17 illustrates how poppets are actuated to create the various logic functions. There are two types of poppets: the double internal and the double external. Depending on how they are employed, they can create passive or active logic devices.

Passive Devices

The OR device. Consists of a double internal poppet actuated directly by the input signals. The arrangement is also known as a shuttle valve and is a popular device in standard pneumatic and hydraulic circuits. Input signals to a OR b (or both) pass directly to the output. If only one of the two inputs is receiving a signal, the poppet action seals the inactive input so that no false signal is created to pass backward through the interconnected logic devices.

The AND device. Consists of a double external poppet actuated directly by the input signals. If only one input a OR b is actuated the poppet seals the passage to the output. If a AND b are both receiving input signals then the stronger of the two signals will seat one port and allow input to the other to pass through to the output port. An output signal is created only when the a AND b signal requirements are satisfied.

The movable-part logic devices AND and OR described above are insensitive to variations in input signal levels. This is in contrast to the jet deflection devices described earlier, which required good signal level regulation for reliable operation.

Active Devices. The poppets employed in the active devices are the same as the passive devices except that they are actuated by a piston or diaphragm. This configuration provides two additional logic functions.

The NOT and INHIBITION device. If input a is receiving a signal and input port b is NOT, input a will pass through the device to output port U.

$$U = a \text{ AND } (\text{NOT } b) = a\bar{b} \rightarrow \text{INHIBITION}$$

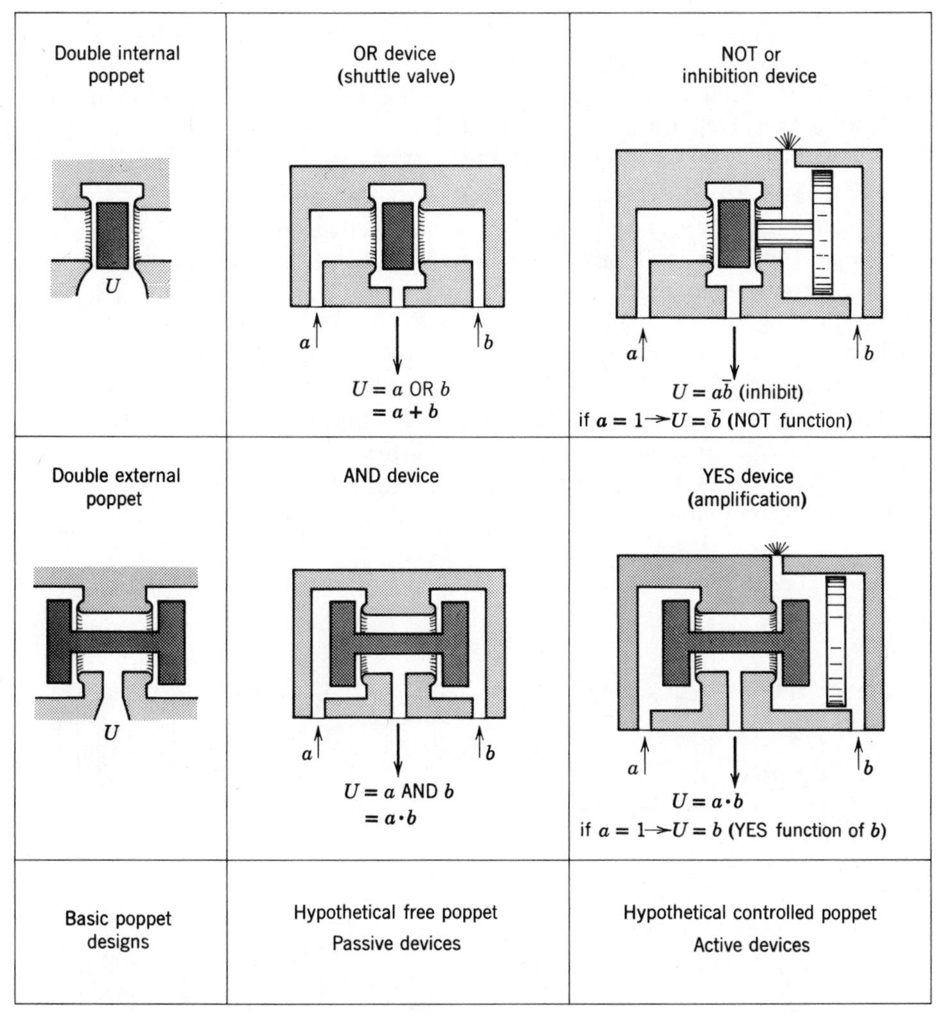

FIG. 3–17. Principles and hypothetical examples of poppet logic devices.

If input b receives an input after input a, the diameter of the poppet actuating piston in port b enables input signal b to overcome the input a signal, thereby switching off the output at U. In this position the output can return through U and exhaust to atmosphere from a vent port in the device.

The NOT function is created when supply is maintained on input a

$$a = 1 \rightarrow U = \text{NOT } b = \bar{b}.$$

The YES device (amplification). When both inputs (a and b) are receiving signals, input a will pass through to output U.

$$U = a \text{ AND } b = ab \rightarrow \text{AND function}$$

This device can be used to create an AND function; however, the passive AND device is usually preferred because of its simpler construction.

If a supply is maintained at input *a*, the device can be used as a YES function. In addition the device can be used as a regenerator or amplifier. The diaphragm in input *b* allows a low level signal at input to overcome a higher level input at *a*. Depending on the construction of a regeneration or amplification device, the ratio between input and output signal levels can reach 1/1000.

Figure 3-18 illustrates the various types of spool valve configurations used as memories. Later we discuss how spool valves can be used to create all basic logic functions. For now, our interest in spool valves is limited to their ability to create a memory function that is maintained even after a momentary or prolonged lack of air supply. The spool will maintain its last position until an input signal is applied to shift the spool to the opposite side. The only minor exception to this is in the plain packless spool illustrated in the Fig. 3-17. The packless design and a lack of detent allows the spool to shift if the valve is physically turned or subjected to vibration.

The electrical analogy to this maintained pneumatic memory device is the latching type relay.

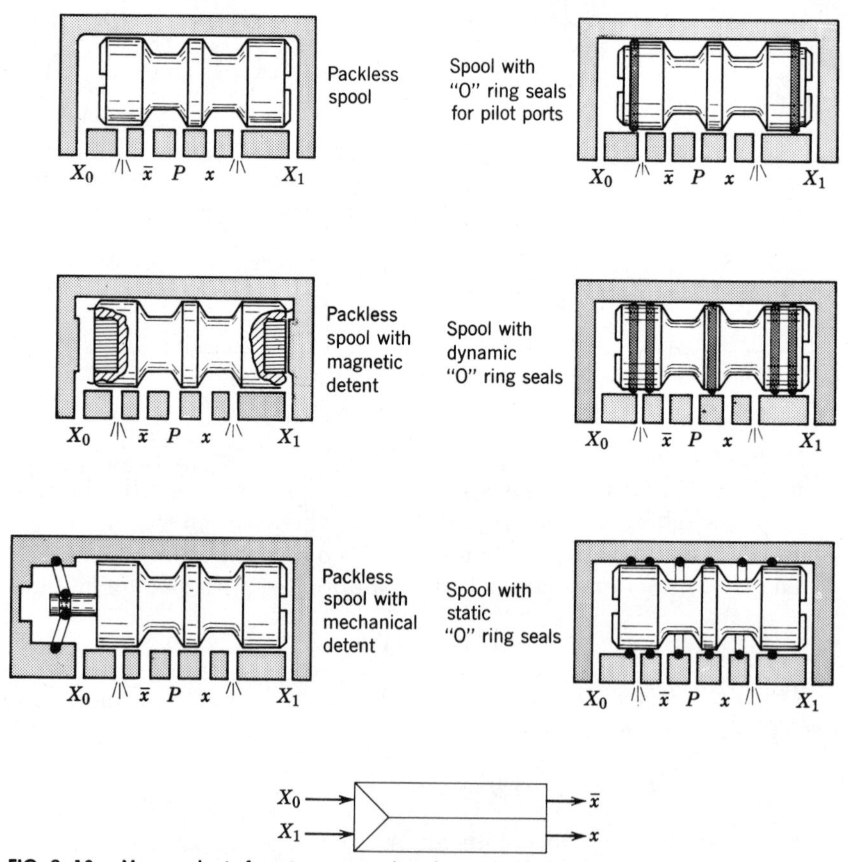

FIG. 3–18. Memory logic functions created with spool valves.

In popular pneumatic valve terminology the memory valve is known as a five-ported four-way double pilot valve. The supply to port P is switched to output ports x or $\bar{x}$, depending on the position of the spool. In either position of the spool, the output port not receiving a supply signal is open to its own exhaust port.

The value of this valve as a memory device is based on its ability to maintain the spool in the desired position; therefore, a mechanical or magnetic detent or friction seal is usually emloyed.

3-4-3 Design and Application of Movable-Part Devices

3-4-3-1 Low-Pressure Movable-Part Devices

In general these devices consist of a diaphragm which actuates a poppet. Figure 3-19 illustrates two types of devices, both of them equivalent to three-way spring-return valves.

The first device consists of a lever which alternates between the interchangeable supply and exhaust ports. The lever is actuated on one side by a diaphragm which opens a passage between two ports. When the diaphragm is deactuated a spring returns the lever to block one passage and opens another.

The second device consists of two linked diaphragms which control the switching action in the same manner as the double external poppet device. The spring pressing against one of the diaphragms holds the passage open between the two of the ports. An input signal against the other diaphragm causes the exhaust to be blocked and the passage between the other to open.

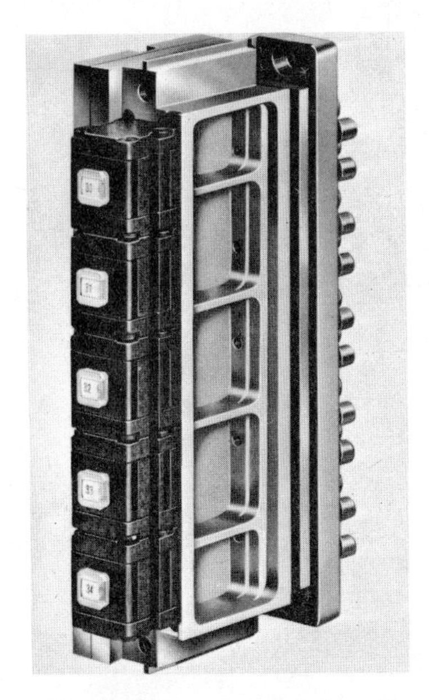

Low-medium pressure logic modules. (Courtesy Samson, Federal Republic of Germany and General Fluidics, USA.) These five units operate according to the principles illustrate in Fig.3-19. Each diaphragm-spring element occupies slightly less than 1 in.^{3} of volume space and attaches to base plates that assemble on hangers to form logic panels.

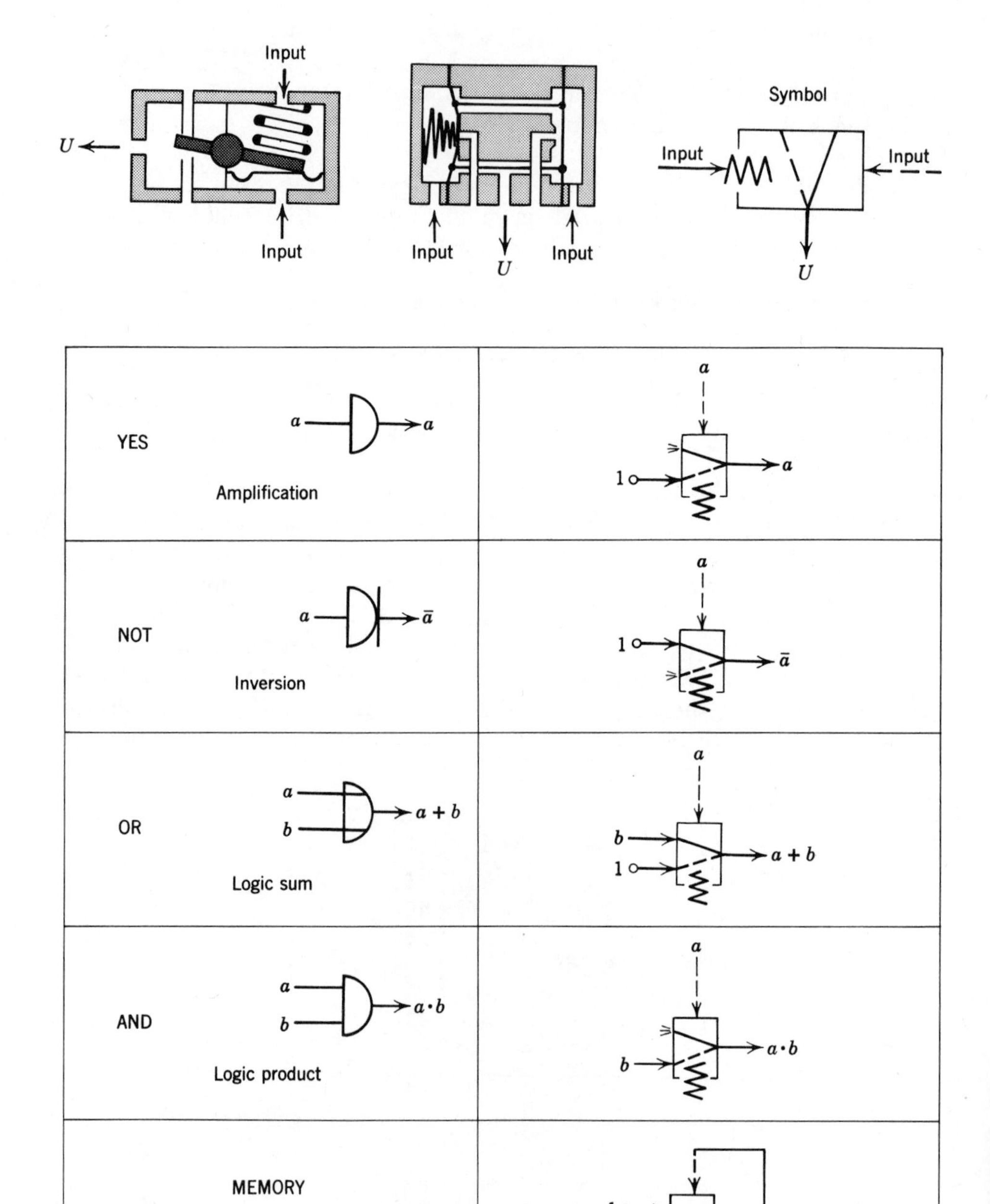

FIG. 3–19. Two types of multifunction diaphragm-poppet devices. Both of them have equivalent operation to a three-way spring return valve.

As illustrated in Fig. 3-19, both devices are capable of creating all the basic logic functions. The versatility of these devices provides the basis for constructing complete logic systems that are able to solve all logic problems.

The drawings illustrate the manner in which the inputs, outputs, and supply are connected to create the basic functions. Note that the memory function requires two devices. Since we are now dealing with spring-return devices instead of spool valves, we must provide a "holding circuit" to maintain either of the two memory states X and $\overline{X}$. The key to the holding circuit is the feedback from the outputs to maintain the opposite output in the "off" position. This method of creating the memory function operates satisfactorily as long as the control system supply pressure is maintained. The disadvantage of this method becomes obvious when the supply is temporarily removed—the memory is lost; this eliminates the possibility of starting the control system at the position on the sequence at which it was stopped.

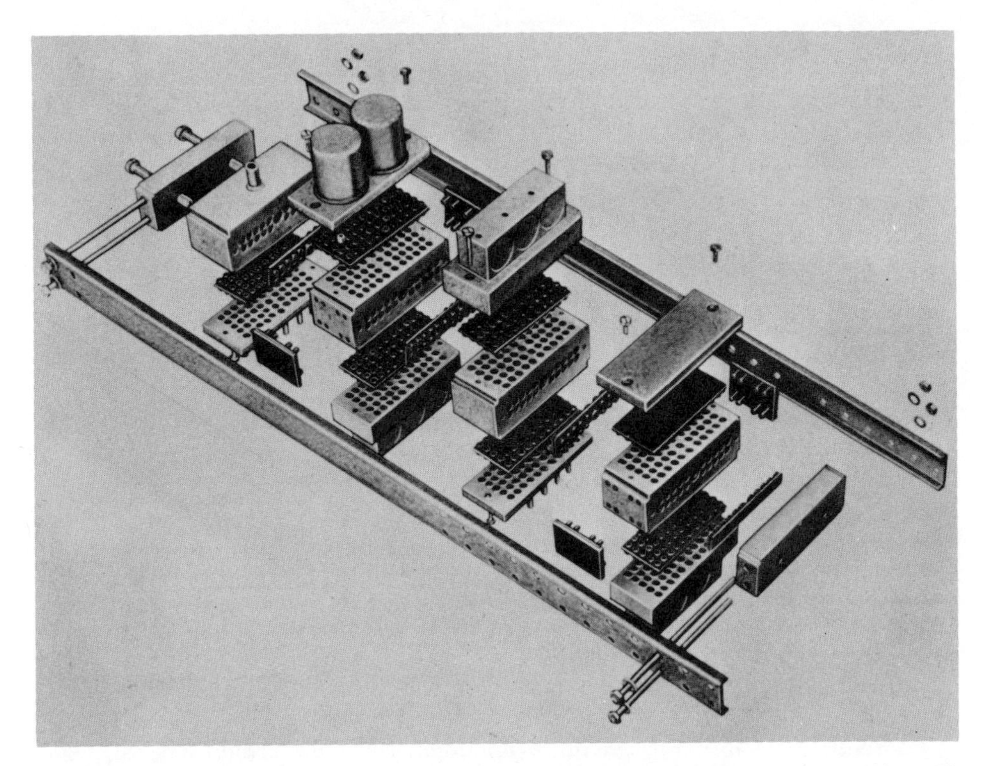

Low-pressure movable-part system. (Courtesy VEB Reglerwerke, German Democratic Republic.) The baseplates onto which logic modules and peripheral components mount are held in place by networks of profiled rails. Circuit connections take place within the baseplates and gaskets seal off leaks. Poised above major baseplates are one module with two chambers for timing volume, a logic module with three elements within, and a sealing plate. Positioned just below these three are another three element logic module, a connector plate, and a two element logic module. The fundamental principles of this system are described in Fig. 3-20. This system is known as DRELOBA.

Figure 3-20 illustrates a low-pressure movable-part system which employs two devices to create the basic logic functions.

- The OR device uses a single poppet made of rubber.
- A multifunctional device is used to create all other functions. This device consists of an external double poppet and diaphragm assembly, which responds to the input signals. The creation of the various functions YES, NOT, and AND are possible by employing the illustrated connections to the inputs, outputs, and supply.
- The memory function is created in the same manner as in the previous device—with holding circuits.

Low-pressure moveable part system. (Courtesy Computer and Automation Institute, Hungarian Academy of Sciences.) The majority of modules in this view constitute hardware of the basic concept described in Fig. 3-21, from the system called USEPPA, designed at the Institute of Control Sciences in the USSR. The exception are three devices in the background with circles on their frontside which are first generation elements from the Hungarian system TRIMELOG.

Figure 3-21 illustrates a multifunctional three-diaphragm type of logic device. Combinations of various levels of input and supply pressures actuates the diaphragms which are interconnected by a linkage that creates a poppet action at each end of the device. This double poppet action provides either a signal at the output port or connects the output port to the exhaust port.

Three different levels of supply pressures are employed in this device: P = nominal pressure: $P_1 = 0.3P$; $P_2 = 0.7P$.

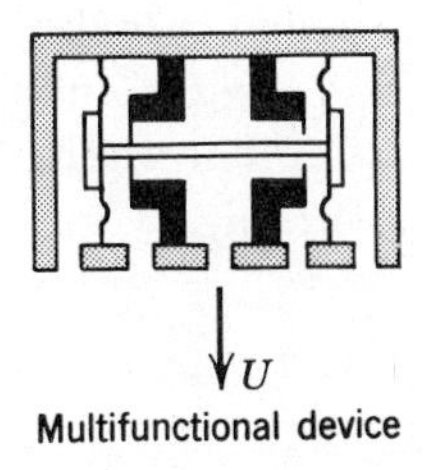

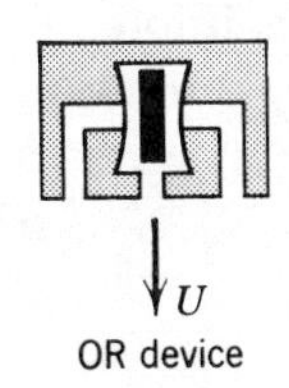

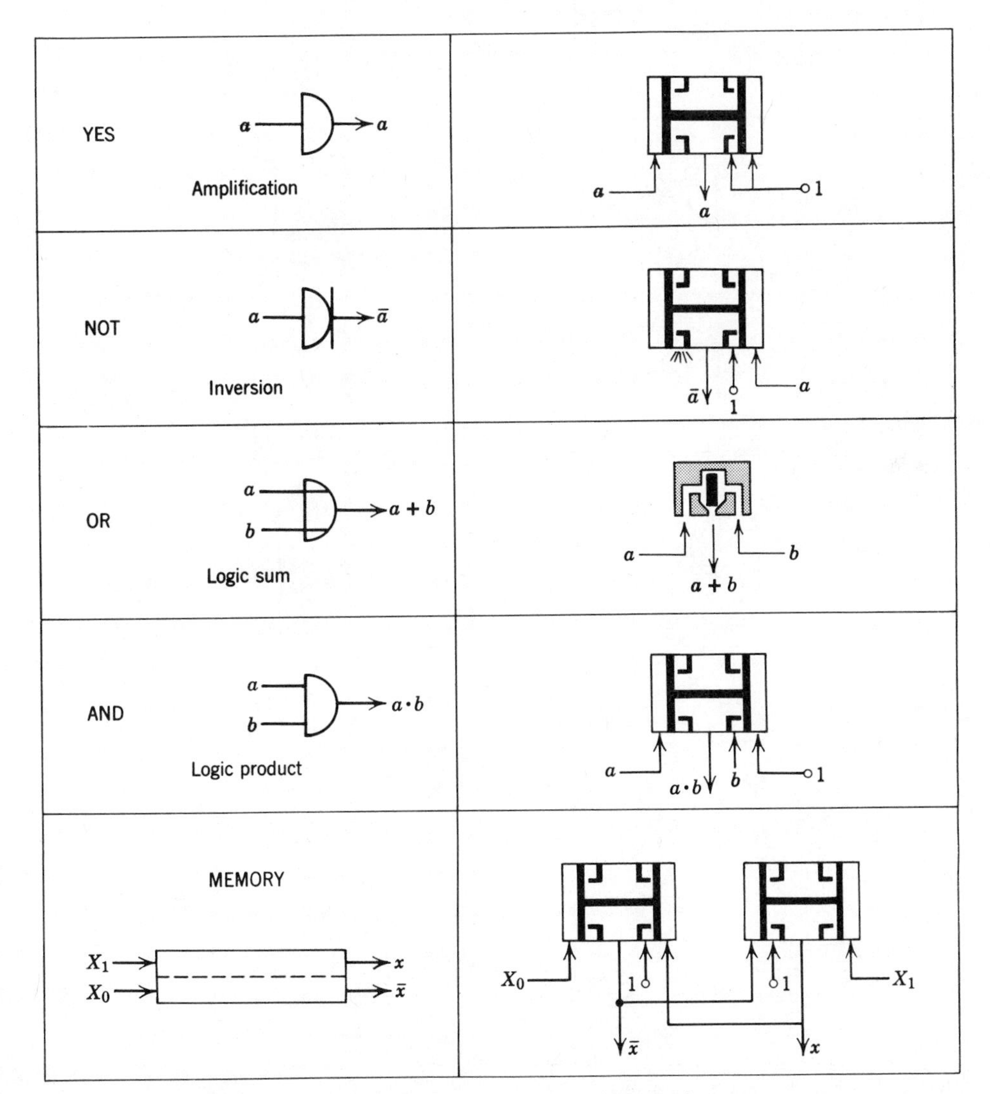

FIG. 3–20. A logic system employing two low-pressure movable-part devices.

The area of the middle diaphragm is twice the size of the diaphragms on each end. The implementation of the basic logic functions becomes apparent from an examination of the illustration. The memory function is created by combining two devices into a holding circuit.

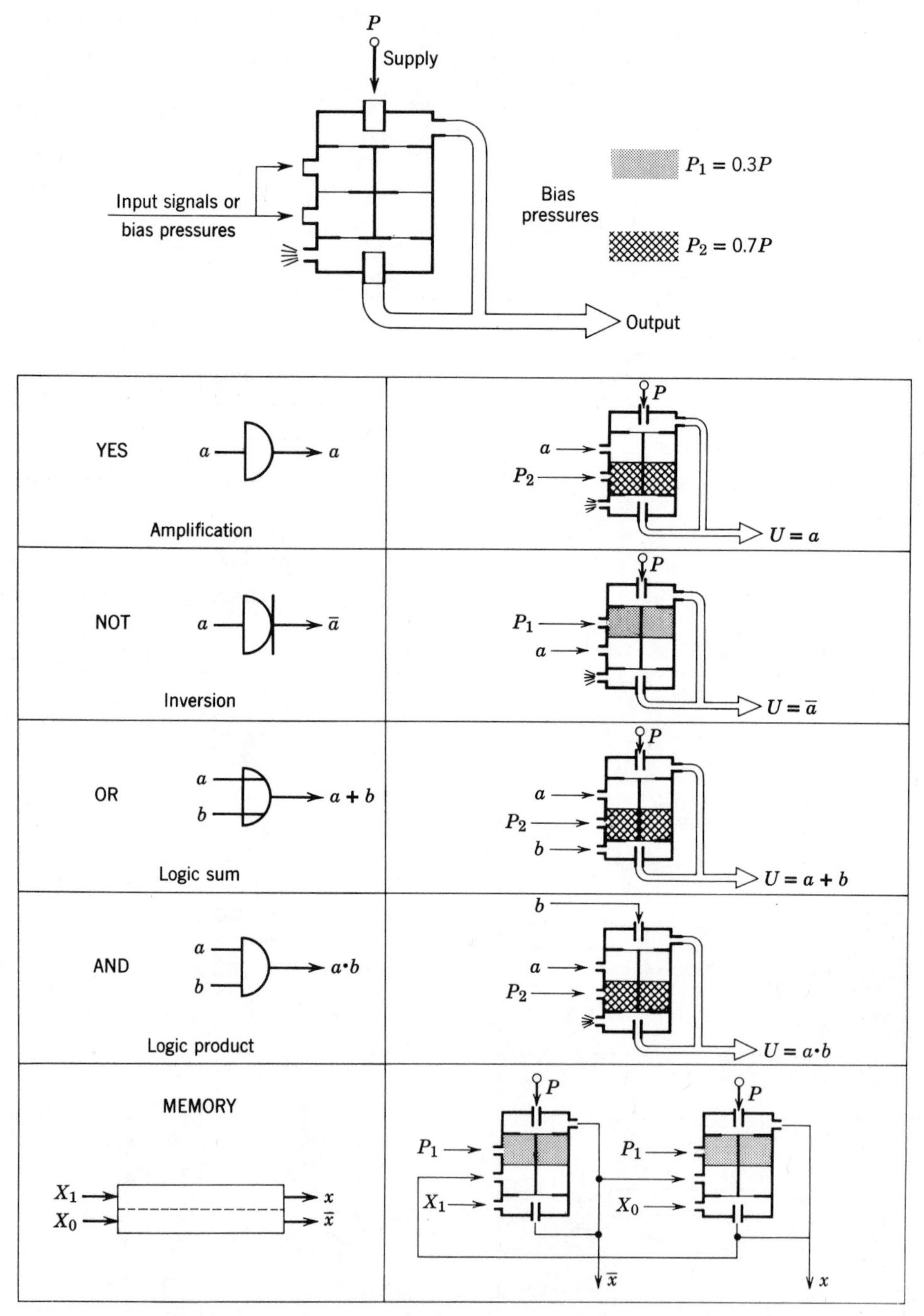

FIG. 3–21. Multifunction three diaphragm device.

Figure 3-22 illustrates a low-pressure movable-part logic system which employs three devices to create the basic logic functions, the principles of which have previously been explained in Fig. 3-17.

The OR function is created by a passive, single poppet device.

The YES and the AND functions are both created by an active, double external poppet device.

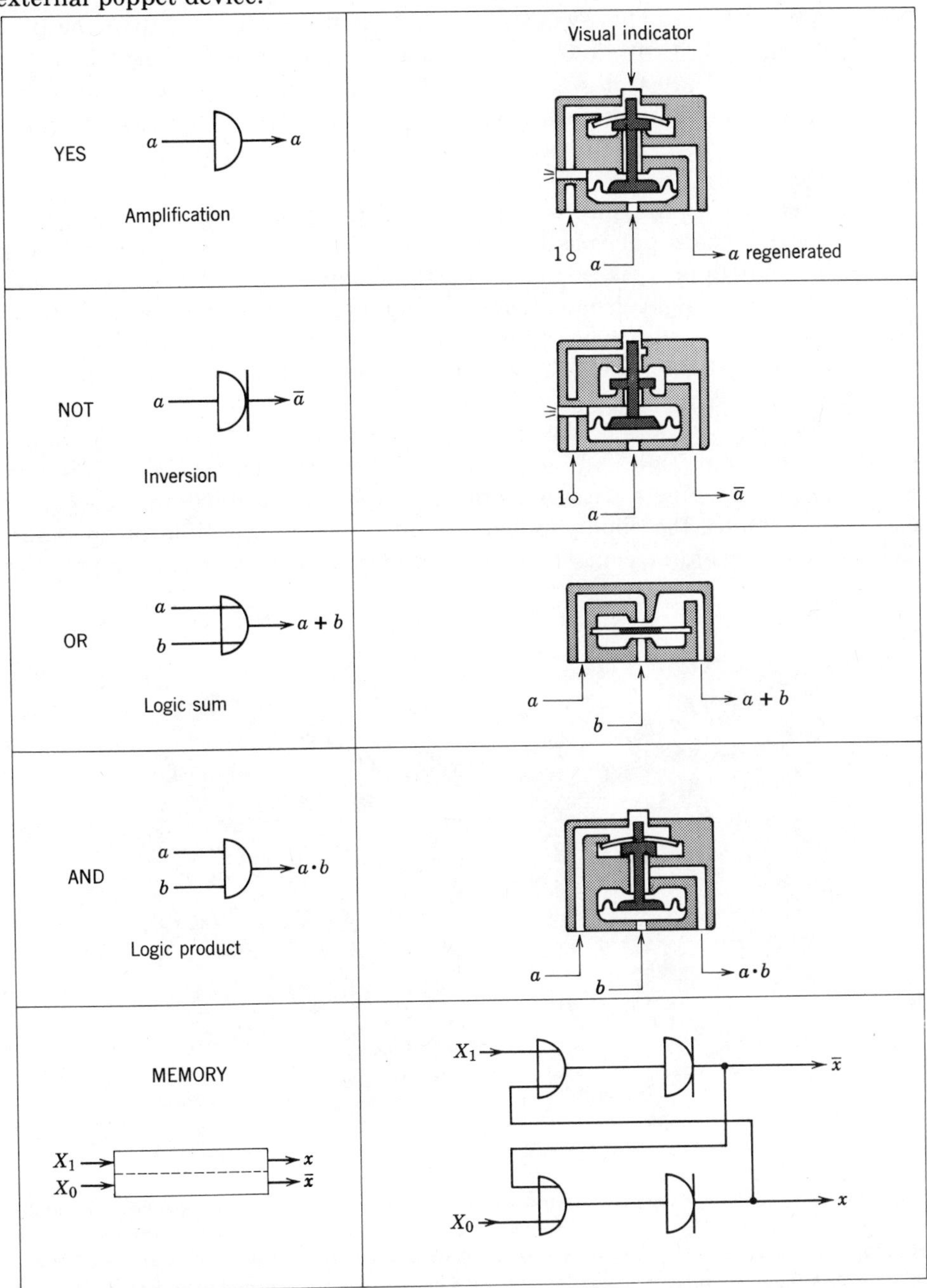

FIG. 3–22. Very low-pressure movable-part logic system.

The NOT function is created by an active, double internal poppet device.

The memory function is implemented by a NOR circuit.

As illustrated in the figure, the two active devices are equipped with built-in visual indicators. The indicator is "ON" when the poppet is actuated by an input signal. The entire system is designed to operate on a nominal 3-psig supply pressure. This system of devices is more sensitive than the previously described systems and will operate in a lower range of pressures.

Figure 3-23 illustrates a device that employs diaphragms to control the switching action. This device is based on the universal NOR function principle. It offers three inputs and requires continuous venting to maintain proper operation.

When supply pressure is (ON) and all three of the inputs are not receiving signals, the supply passes through to the output port. Concurrently a portion of the supply passes under the three input control diaphragms and down to the lower right exhaust control diaphragm. This diaphragm remains depressed as long as all three inputs are not receiving signals. A vent to atmosphere is also located on the right side to maintain a proper balance within the device.

When one or more of the three inputs are actuated, the flow to the exhaust control diaphragm and vent is stopped. The exhaust control diaphragm snaps upward, opening a passage from the output port through to the exhaust port. Concurrently the supply is backed up from the actuation of an input. This causes the output control diaphragm to snap downward closing off the

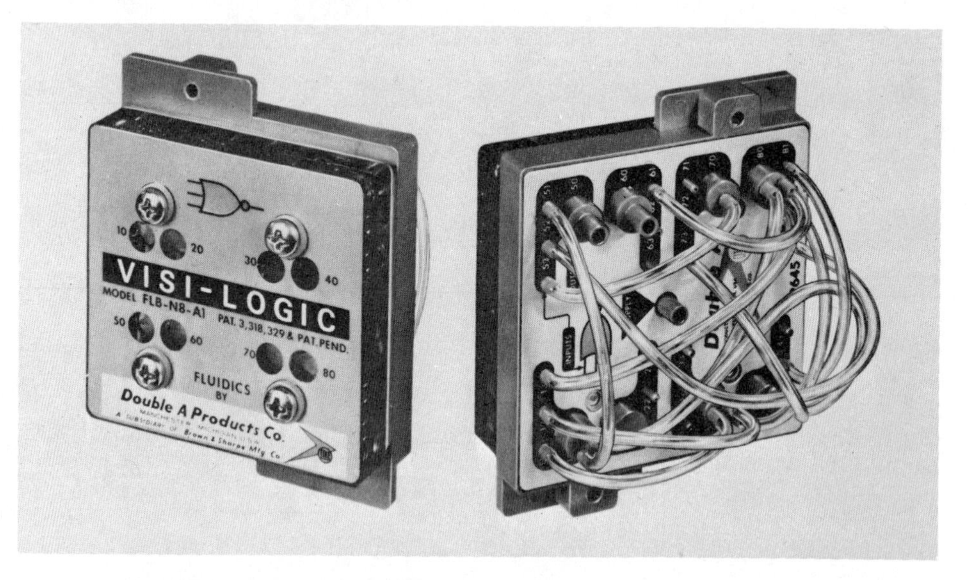

Low-pressure movable-part system. (Courtesy Double A, USA.) This molded plastic sandwich contains eight NOR logic gates of the type illustrated by the principles described in Fig. 3-23. The module on the left is attached to a gasket sealed baseplate with fittings for circuit connections on the backside. One diaphragm of each gate is visible through the eight round windows on the front and these serve the purpose of indicating dynamic action for each logic element.

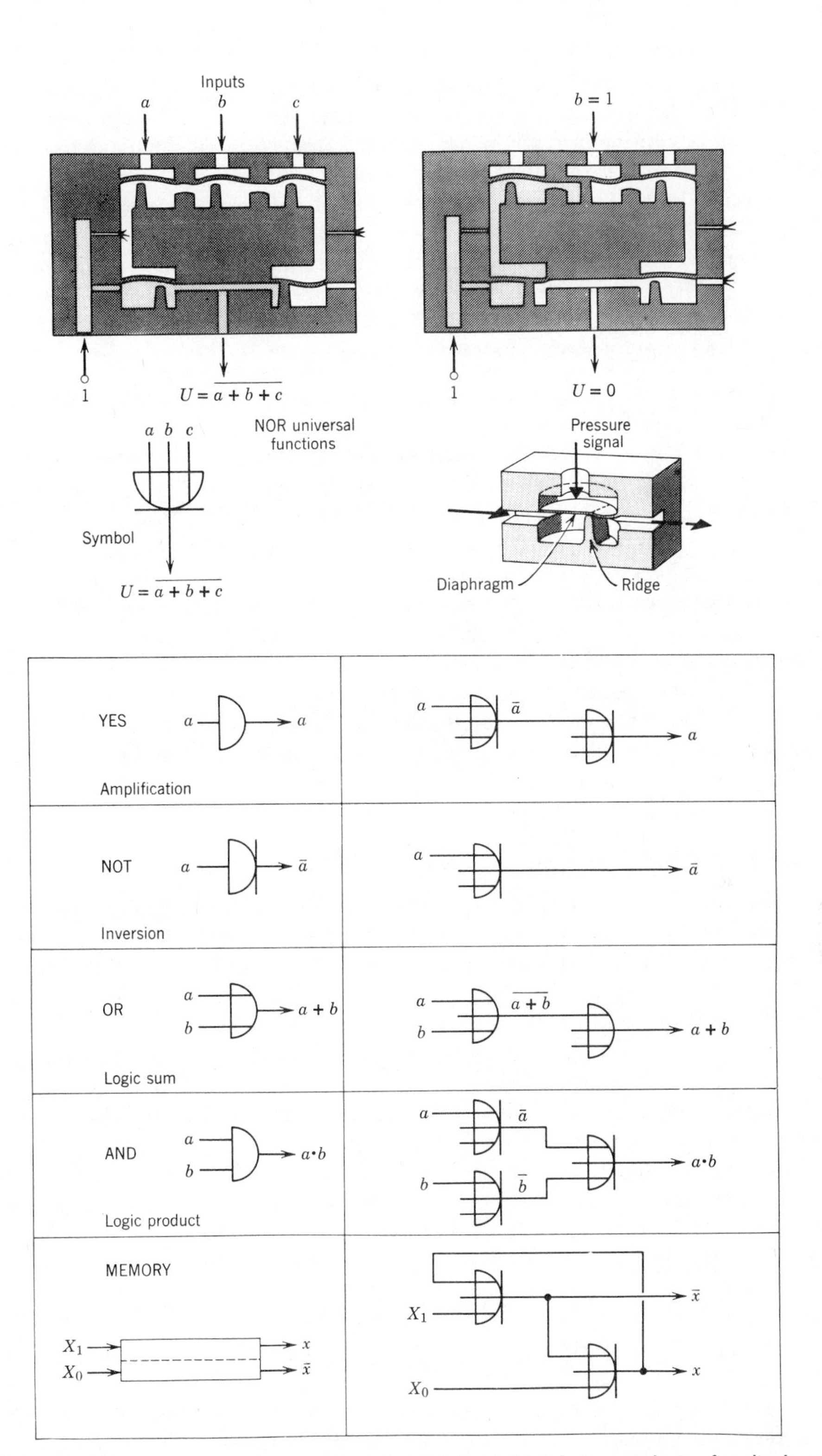

FIG. 3–23. This system is based on a universal type 3 input NOR device, employing five diaphragms and operating on low pressure.

flow to the output port. As illustrated, the device employs five diaphragms which control flow by snapping downward against the ridges when pressure is applied, and snap upward when pressure is removed. This device is unique in that it incorporates some of the basic principles of movable and nonmovable part devices. It requires a continuous venting to atmosphere when the output is "ON." In a similar way to jet destruction devices, it is capable of creating the universal NOR function and all basic logic functions when combined with duplicates of itself as illustrated on Fig. 3-23.

Overall the device is classified as a low-pressure movable-part device. It requires a supply pressure between 1 and 3 psig.

General Characteristics and Construction of Low-Pressure Movable-Part Devices. With the exception of the last device, all low-pressure movable-part devices contain integrated-poppet-diaphragm assemblies. The internal orifices are equivalent to 0.03–0.08 in. in diameter. The inputs respond to pressures as low as 0.5 psig (typical for the last device) to 2 psig. In most cases the diaphragm curtain is damaged when subjected to supply pressures above 30 psig. (The last system described is limited to 3 psig.) The bodies of the devices are usually constructed of aluminium, zinc alloy, or various plastic materials. The diaphragms, usually made of neopreme, are held tautly in place between the body sections.

These devices are usually designed for subbase mounting to facilitate installation and replacement. Small-diameter plastic tubing (typically 1/8 in. OD) is used for the interconnection in conjunction with "push-on" or barbed port terminal connectors.

3-4-3-2 High-Pressure Movable-Part Devices

High-pressure industrial air supplies are usually limited to a maximum of 125 psig. High-pressure logic devices operate within the range 30 to 125 psig. All of these devices are basically miniaturized versions of standard pneumatic power valves. Aside from the size reductions, mounting arrangements have usually been modified to facilitate dense mounting configurations.

Figure 3-24 illustrates a system consisting of miniaturized spool valves. One is a three-way single-pilot spring return which is capable of creating all basic logic functions: YES, NOT, OR, AND, and MEMORY. The memory function would require two valves interconnected to create a holding circuit, as described previously in the section on low-pressure logic devices. However, it is simpler and more advantageous to use a four-way double-piloted spool valve. The memory state will be maintained even when an interruption in supply occurs. In electrical relay terminology this holding feature is referred to as "latching."

Figure 3-25 illustrates two multifunctional devices known as "OR-RELAYS."

Even though the design configurations differ in the two OR-RELAYs, they are almost identical in operation. The only difference is that one uses a spring return and the other uses differential pressure for spool return.

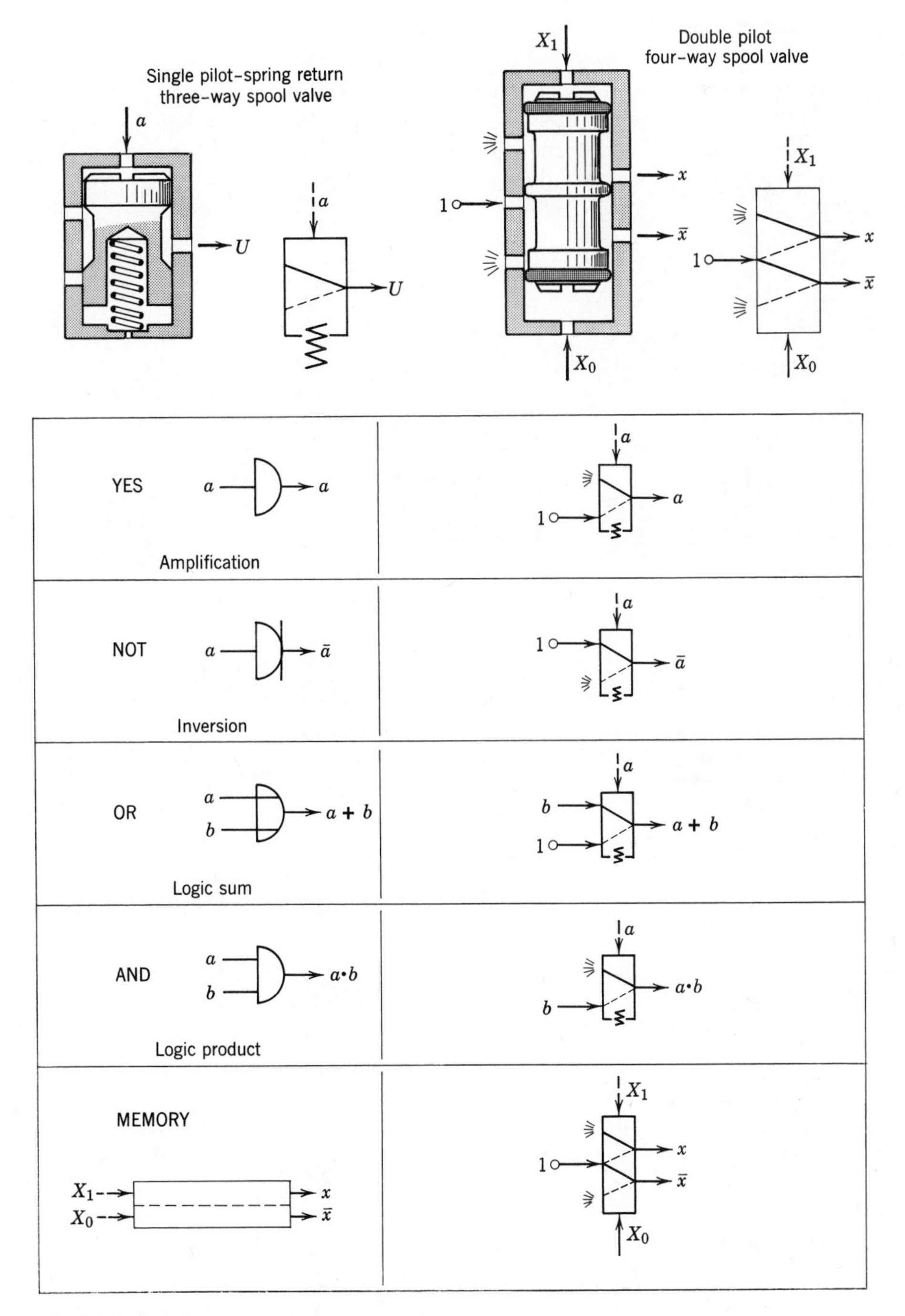

FIG. 3–24. Spool valves employed as logic devices.

A noteworthy feature of these valves is the combination of poppet and spool. This provides two inputs for spool actuation. The symbol for the OR device placed on the top of the valve schematic clarifies this feature. The memory function requires two devices and a holding circuit.

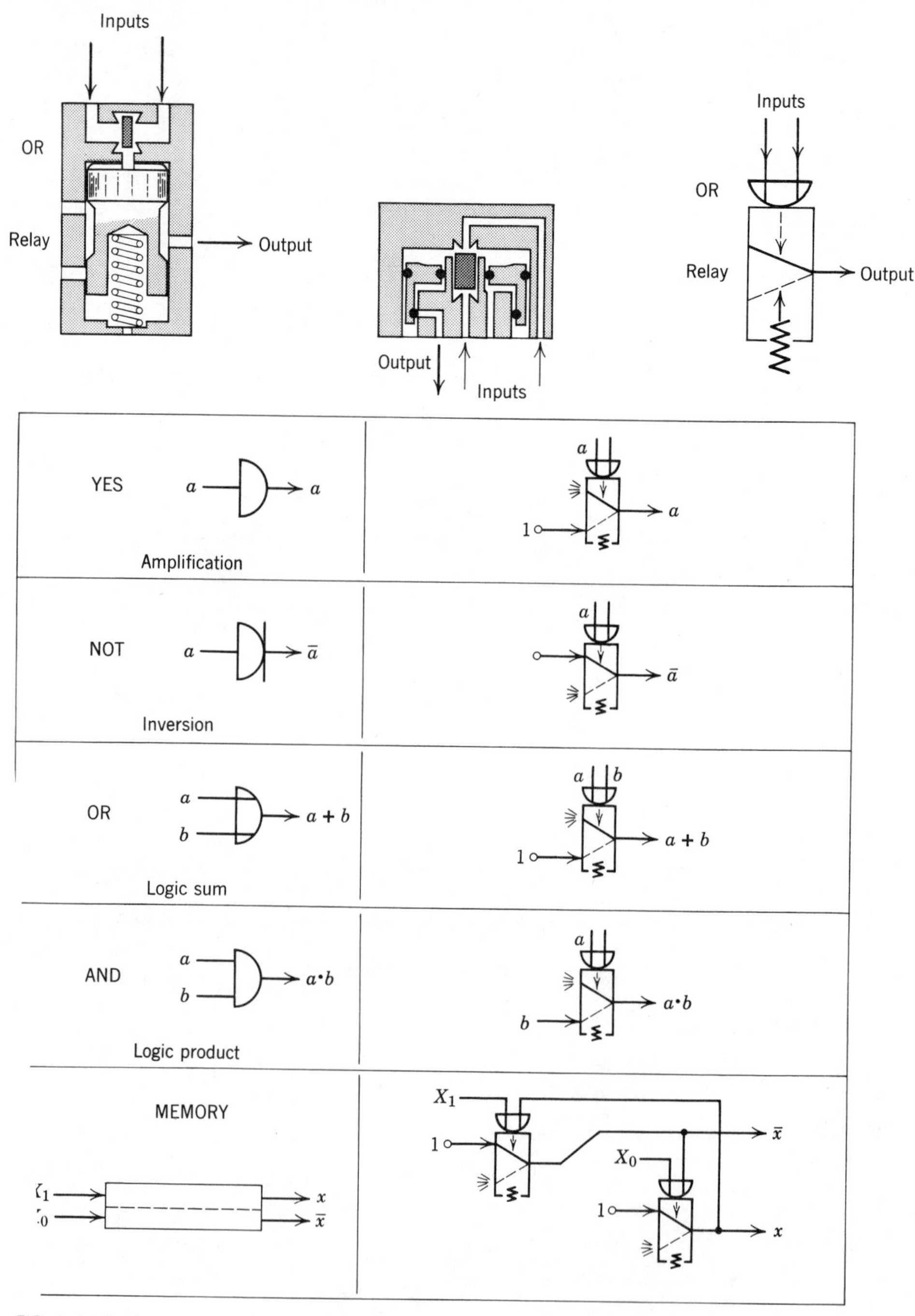

FIG. 3–25. Two types of OR-relay logic devices, capable of creating all basic logic functions.

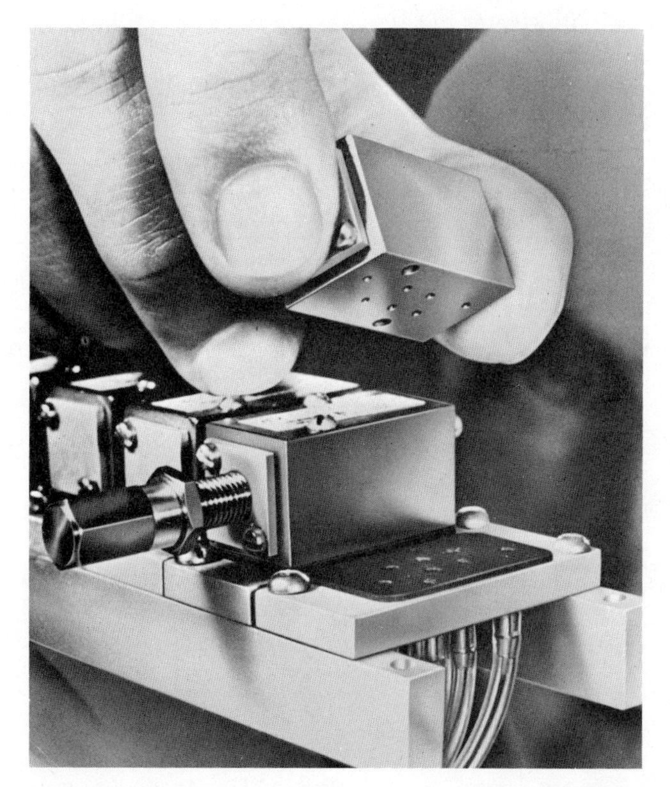

High-pressure movable-part system. (Courtesy Parker Hannifin, USA.) These miniature spool valves seat onto gasket sealed baseplates that have fittings on their underside for making circuit connections. The basic principles of this system are illustrated in Fig. 3-24.

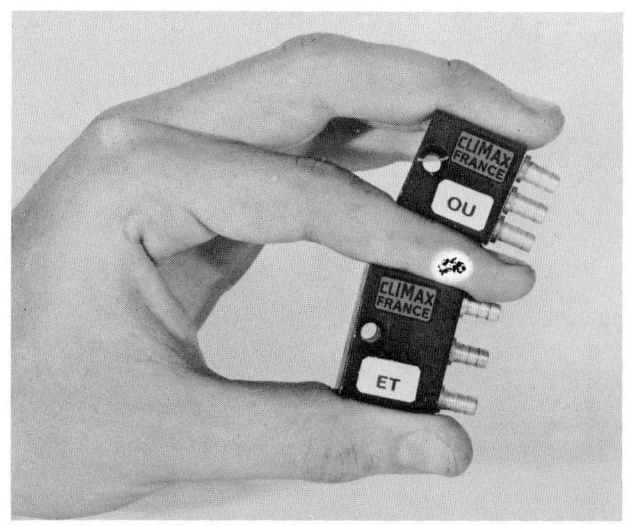

High-pressure free poppet devices. (Courtesy Climax-France, France.) To make connections, plastic tubing is simply pushed onto the barbed fittings shown as part of these AND and OR modules.

Figure 3-26 illustrates two design variations on inhibition logic devices. The basic operating principle of these devices was described in Fig. 3-17. The output U is "on" if input b is "on" AND input a is NOT.

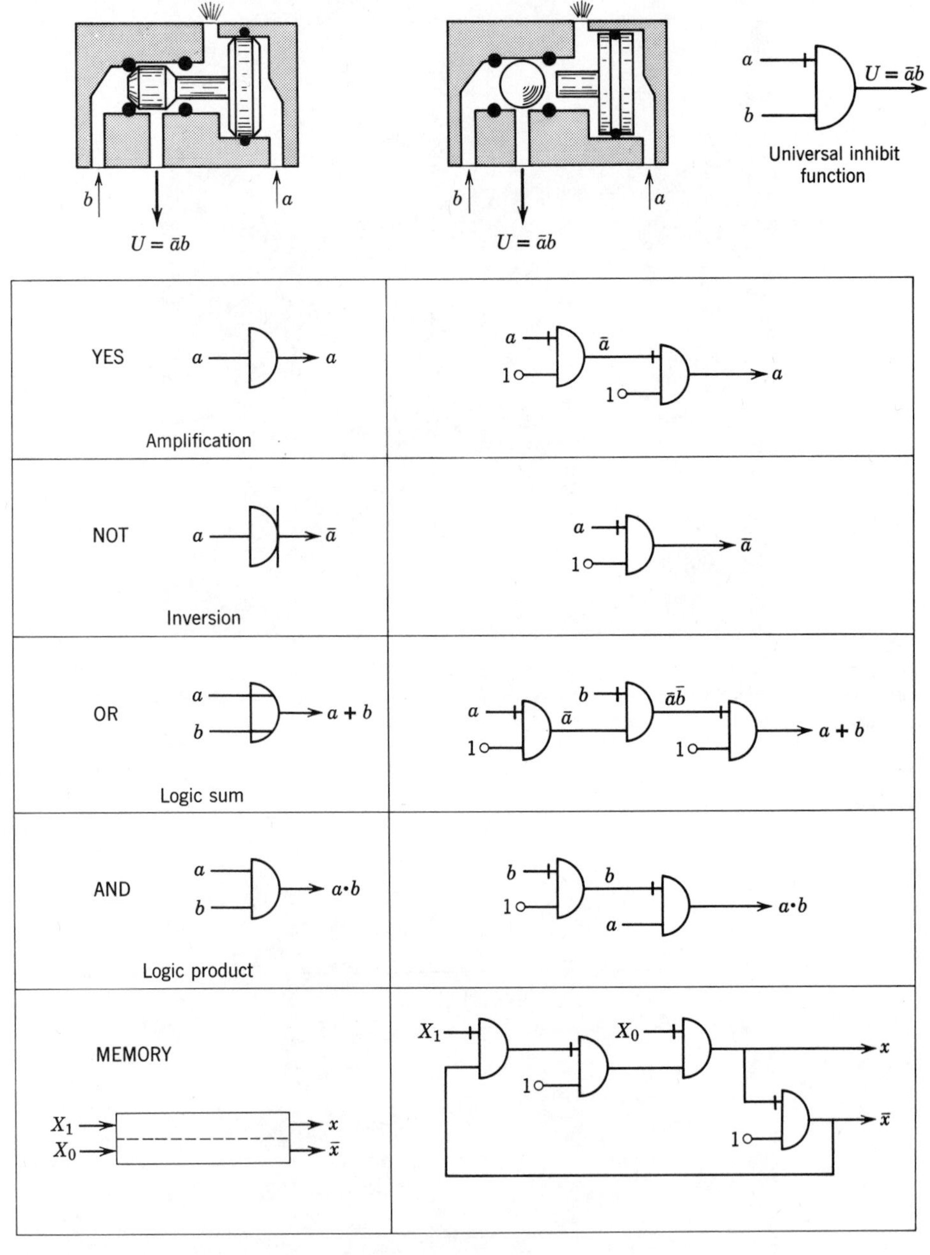

FIG. 3–26. Two universal type devices performing the inhibit logic function.

$$U = b \text{ AND } (\text{NOT } a) = b \cdot \bar{a} \rightarrow \text{inhibition function}$$

All basic logic functions are possible with the inhibition device, since the inhibition function is universal. This means that all logic problems can be solved. The logic function chart in Fig. 3-26 points out the need for multiple devices to create all but the NOT function. As a result of this requirement, it was obvious that a complete system would necessitate a large number of these devices.

High-pressure movable-part system. (Courtesy Clippard, USA.) This group of high-pressure movable-part devices plug into octal sockets in the subbase. The basic logic device is a spool valve and those elements with knobs on their tops are flow control and/or pressure regulator elements while others show spool indicator windows. The basic injection molded body may contain 26 different valve and control functions to fit every circuit need.

Figure 3-27 illustrates a system similar to the previous one except that two different devices are required and more importantly only one of each device is required to create each of the logic functions, except for the memory function.

- The versatility of the multifunctional device is due in part to the various cross-sectional areas of the spool.
- The OR device uses the principle of a passive poppet which is employed universally as a standard shuttle valve in pneumatic circuits.

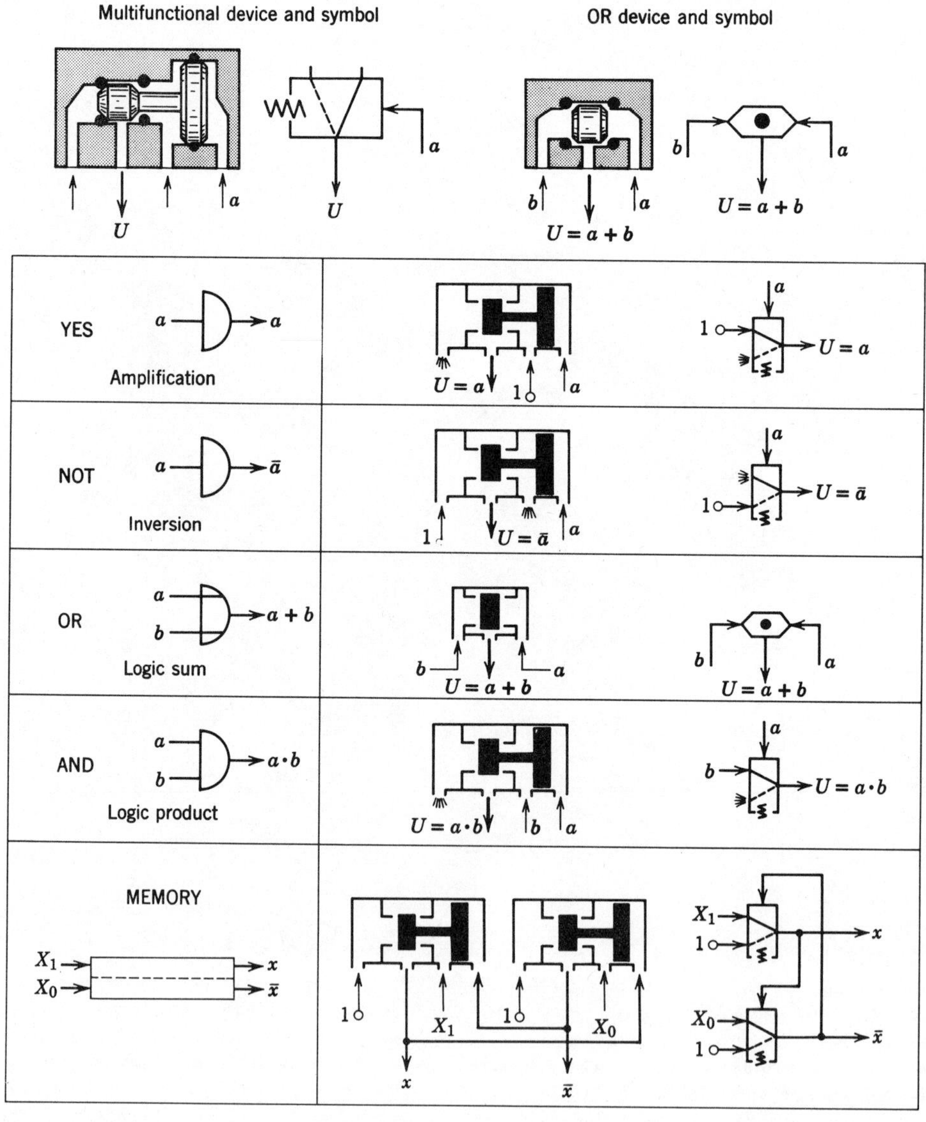

FIG. 3–27. Two high-pressure poppet valves perform the basic logic functions required in a complete system.

High-pressure movable-part system. (Courtesy Jouvenel et Cordier, France.) These modules are mounted on baseplates that attach to the perforated panel. The perforations permit circuit plumbing from behind the panel. One can see AND, OR, NOT, and spool valves for memory clustered toward the topside. Notice the pressure checkpoints on each module. Toward the leftside are peripheral components. From the top down are a solenoid valve, a timer, and an adjustable pressure switch. Then panel plumbing connections are clustered at the bottomside.

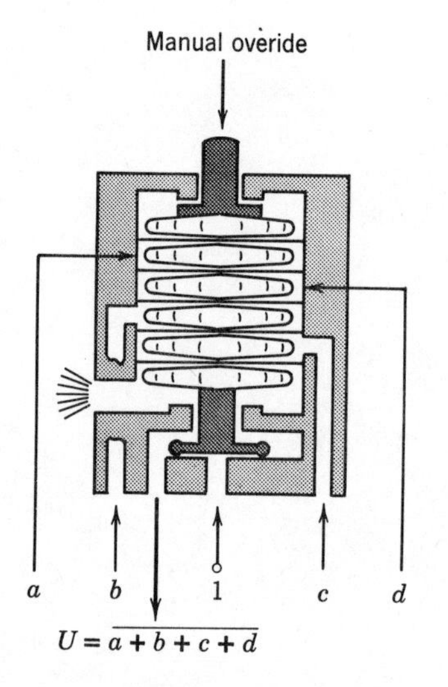

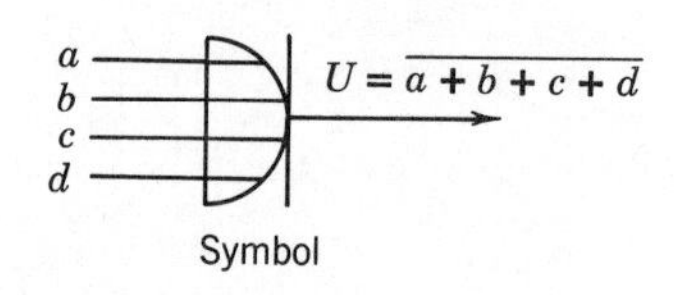

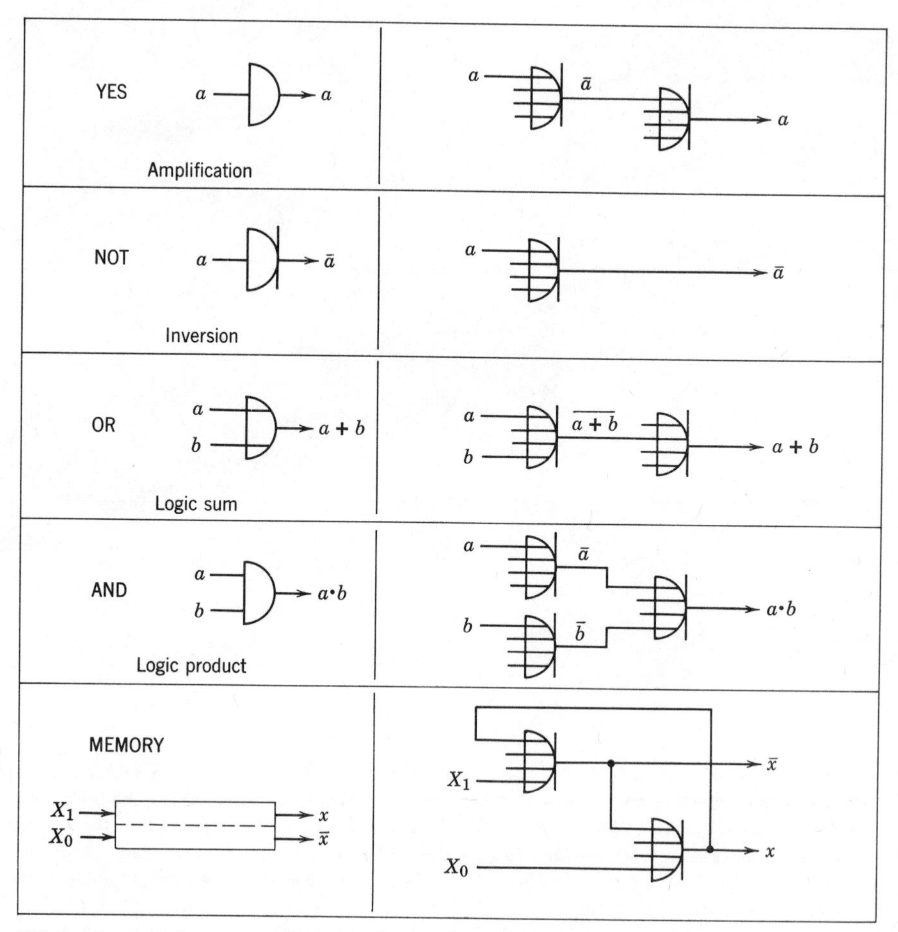

FIG. 3–28. A high-pressure NOR logic device and its application.

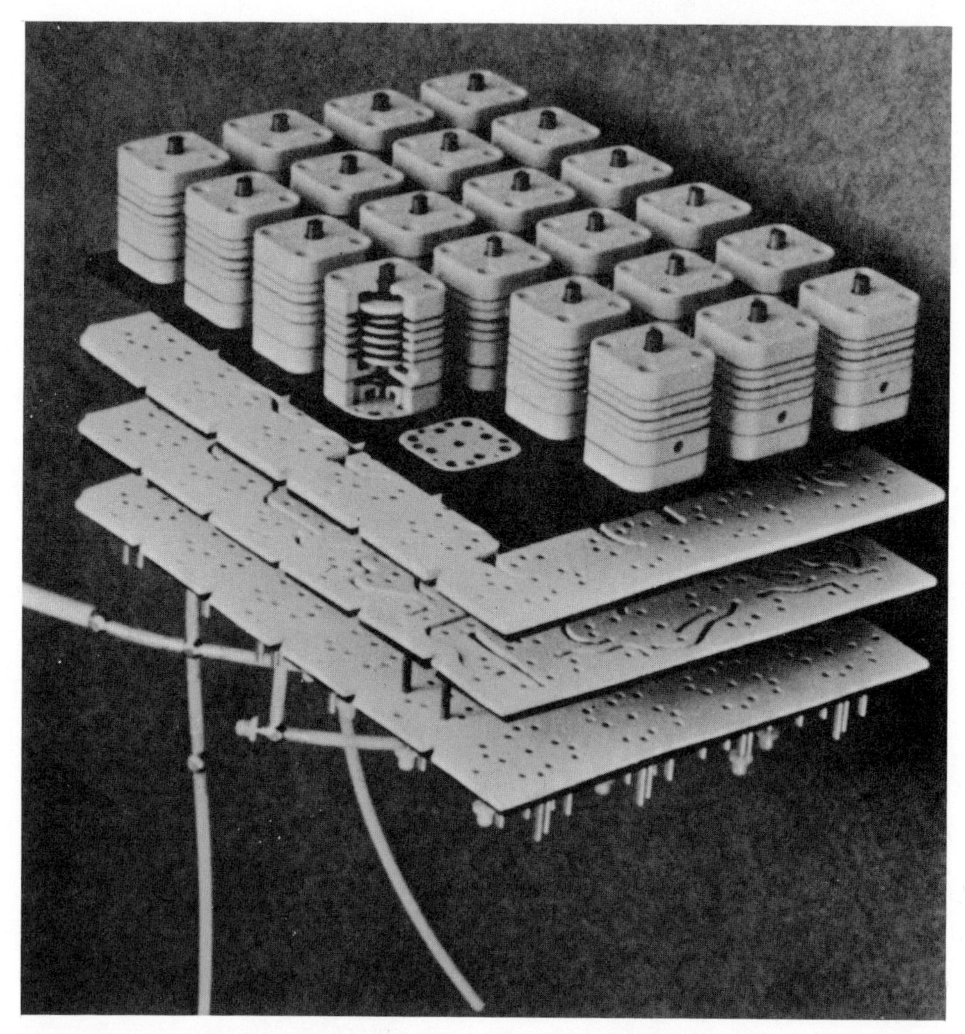

High-pressure movable part system. (Courtesy N.V. Machine-fabriek Sempress, The Netherlands.) These NOR modules are elements of the concept described in Fig. 3-28. The button extending from the top of each is for manual override.

Figure 3-28 illustrates a universal NOR logic device with four inputs. Each input enters a sealed chamber containing one disc. If one or more of these inputs receives a signal, the poppet is actuated by the stack of discs, blocking the supply to output flow and opening a passage between the output and exhaust located in the side wall of the device. A useful feature of this device is the manual override button located on top of the disc. This provides a simple means of manually operating the device for check-out or trouble-shooting.

The application of this device is similar to the jet destruction devices described earlier. To create the basic logic functions no more than two inputs are required. The third input can be useful on ocassion, but the fourth input is seldom necessary.

Figure 3-29 illustrates a system employing three types of poppet valves and a four-way detented spool valve.

The YES and the AND functions are both created with double external poppet devices. An input signal actuates the diaphragm, which in turn actuates the poppet. A spring return is used to reset the device when the input signal is removed.

The NOT function is created with a double internal poppet design. The OR function is created with a passive OR device; an internal diaphragm flexes to allow input signals from a OR b to pass through to the output.

The memory function is created by a four-way spool valve with a mechanical detent. As mentioned previously, this has the advantage of retaining its memory even after an interruption in the air supply.

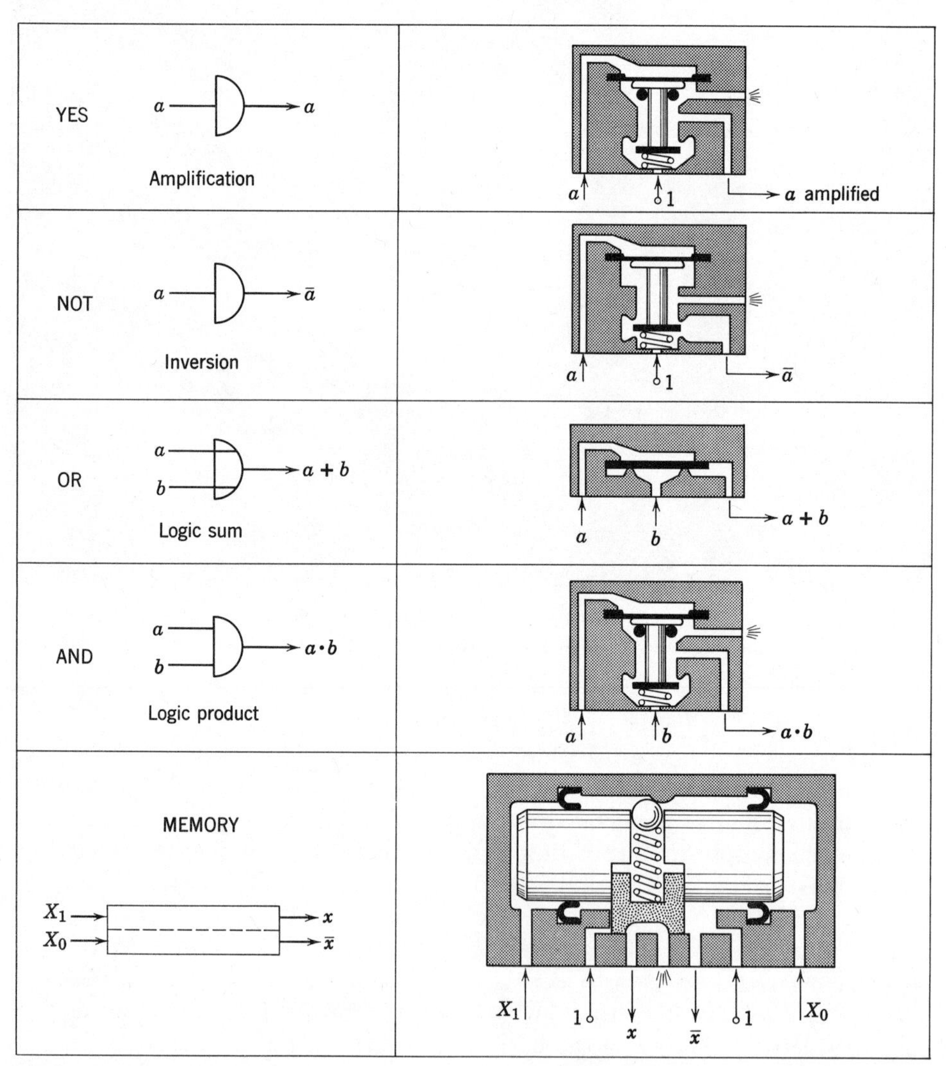

FIG. 3–29. A high pressure logic system employing three types of devices: a poppet, poppet and diaphragm, and a detented spool valve.

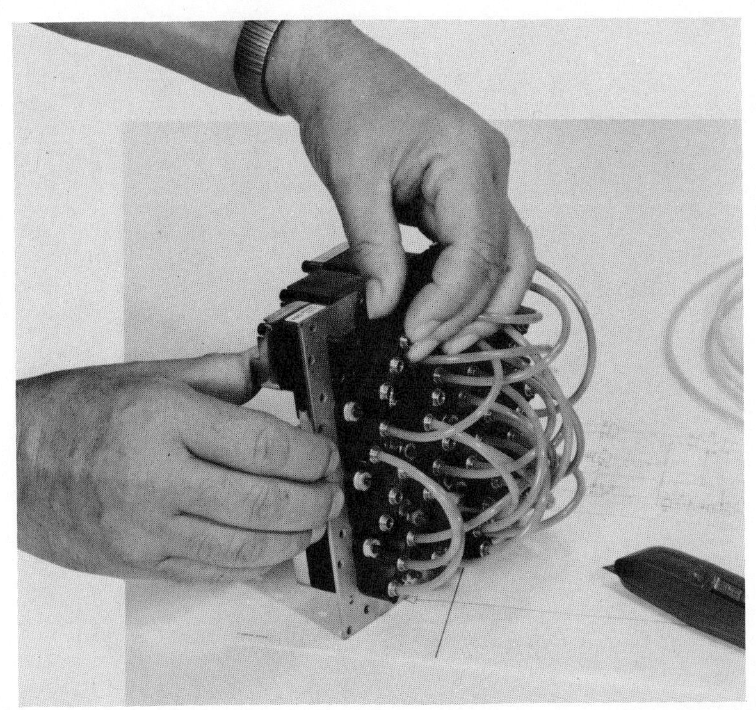

High-pressure movable-part system. (Courtesy ARO, USA.) The various logic modules, some of which are illustrated in Fig. 3-29, attach and are O-ring sealed to a baseplate. In the top view, this baseplate is laminated and contains the circuit connecting channels. The view below shows plug-in connectors where circuit plumbing offers interconnecting versatility.

Figures 3-30 and 3-31 extend the idea of multiple types of devices. In these two systems each device is individualized for each logic function. The operating principles for each valve have been discussed earlier:

- YES—diaphragm actuated external double poppet.
- NOT—diaphragm actuated internal double poppet.
- OR—internal double poppet.
- AND—external double poppet.
- MEMORY—four-way detented spool valve.

Both of these systems have two important practical advantages:

- Only one device is necessary for each function, including memory.
- Each device can be used only one way; this simplifies tubing connections and reduces chances of tubing errors.

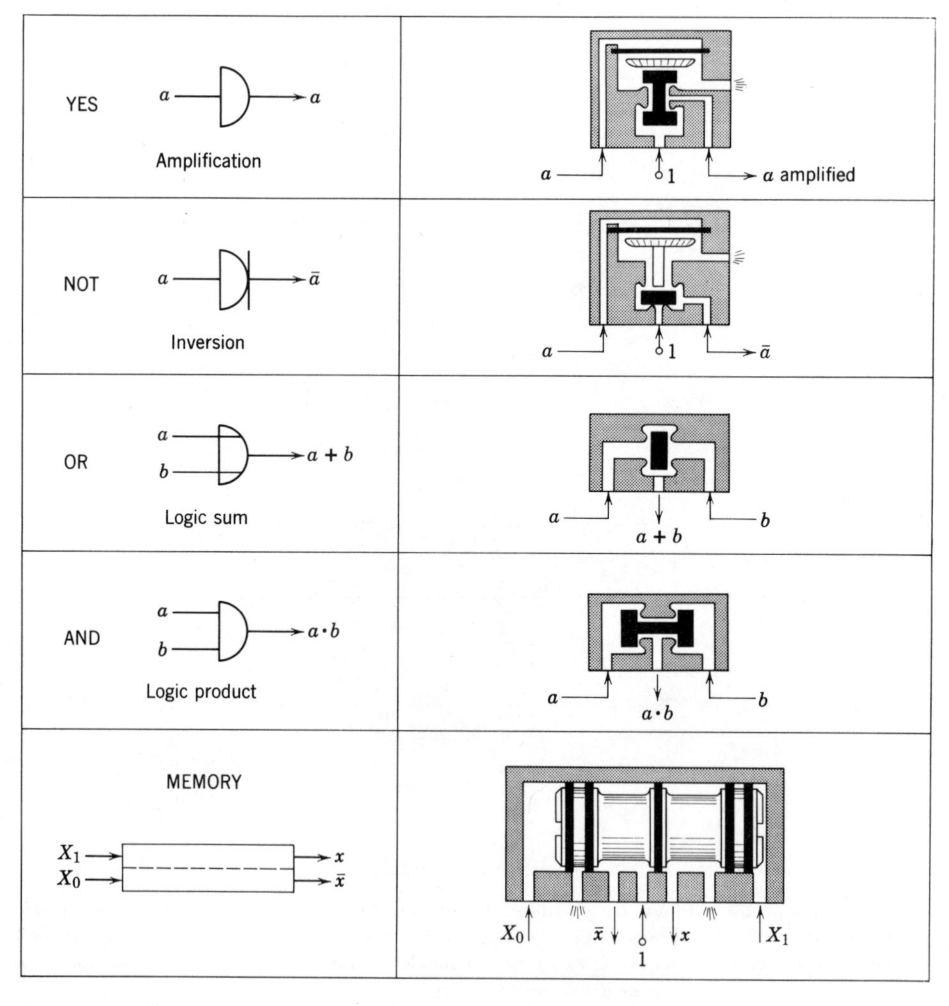

FIG. 3-30. A high-pressure logic system employing individualized devices for each function.

Figure 3-31 illustrates a system employing individualized devices for each function, pressure indicators on each device, and manual overrides on the YES, NOT, and MEMORY.

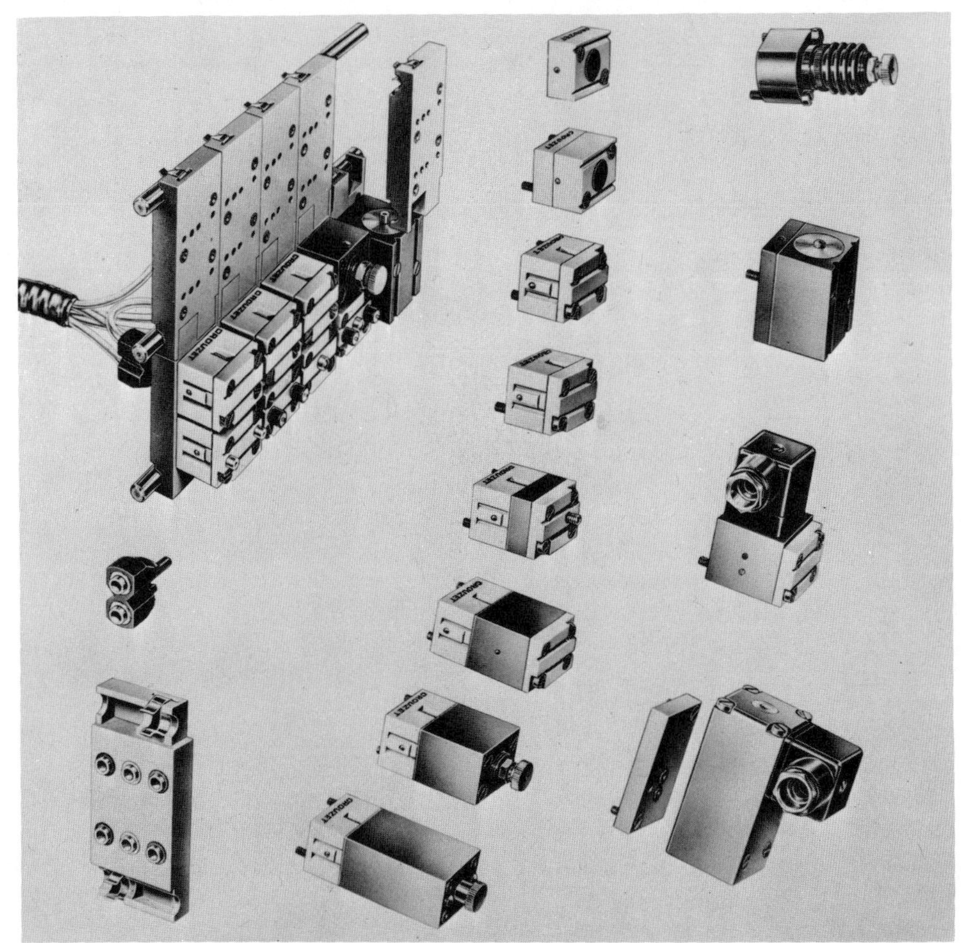

High-pressure movable-part system. (Courtesy Miller Fluid Power, USA and Crouzet, France.) The baseplates onto which logic devices mount are shown attached to round hangers with integral clips, thus assembled into panels. Each baseplate contains integral plug-in connectors so that circuit plumbing accumulates behind the finished panel. What is visible are OR, AND, YES, and NOT modules, a pressure amplifier, an impulse generator, an adjustable pressure switch, and a timer; all centered in the view from the top down. On the right from the top are a pressure regulator, a spool valve memory element, a pressure-electric switch, and a solenoid valve. All of these parts are designed around the principles described in Fig. 3-31.

The pressure indicators are small pistons designed for manual actuation. If pressure is present under the indicator, it will be returned to the extended position. This provides a quick easy way to check or trouble-shoot a circuit. The indicators function is similar to the check light used by electricians.

Construction and Operating Characteristics of High-Pressure Movable-Part Logic Devices. All the devices mentioned operate well on pressures between 30 and 125 psig. Their orifice diameters are equivalent to 0.1–0.2 in.

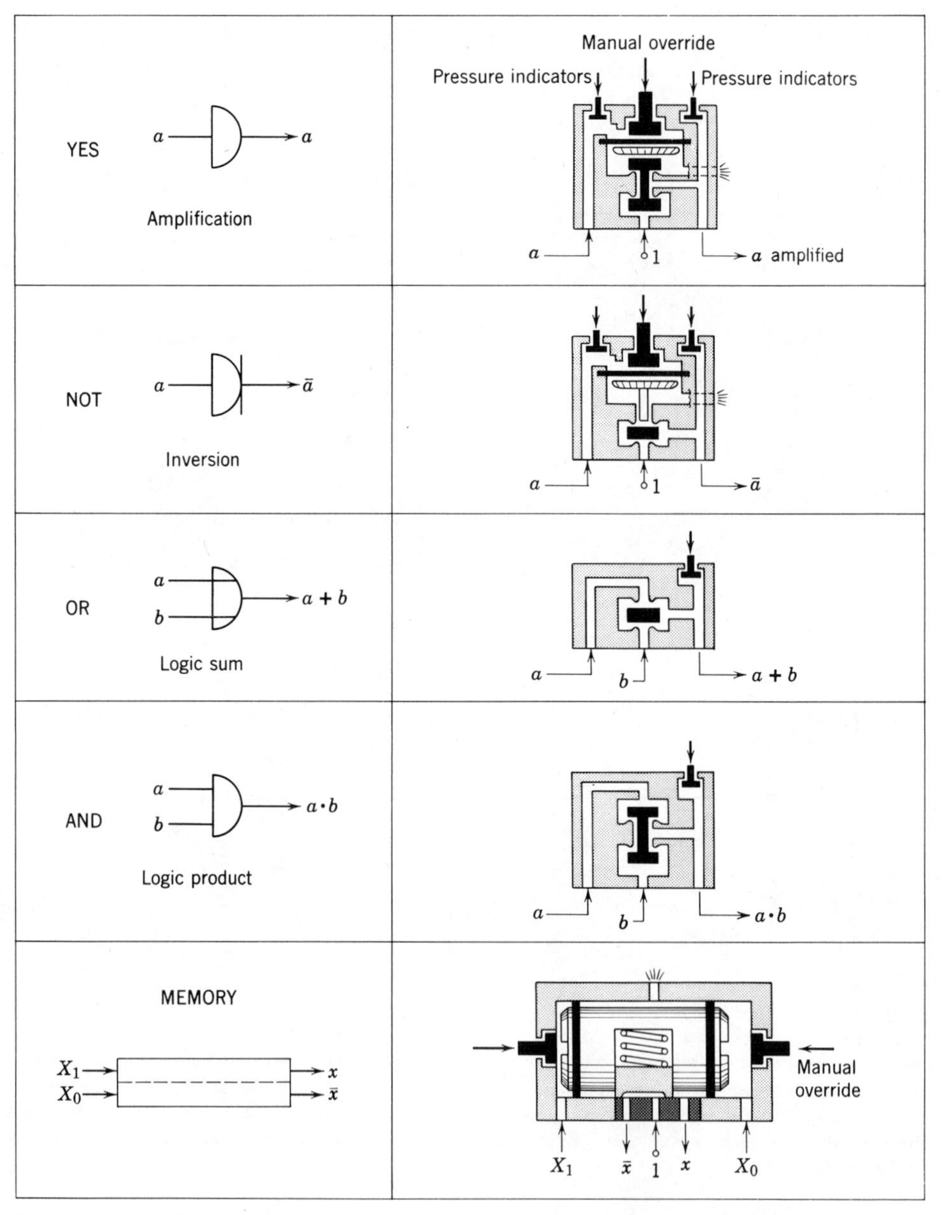

FIG. 3–31. A high-pressure logic system with manual overrides and pressure indicators.

and they are designed for subbase mounting which facilitates installation and replacement. Plastic tubing is used for interconnections.

The bodies, poppets, and spools are fabricated from metal or plastic, and the diaphragms and seals are made from neoprene. Some of the spool valve devices require lubricated air. No lubrication is recommended for poppets, and the trend is to eliminate lubrication on all fluid logic devices.

The response time of the devices is dependent on the stroke length and weight of the moving parts. Poppet devices usually respond to 1/100 sec. or less; spool valves usually require several hundreds of a second.

High-pressure movable-part system. (Courtesy Kuroda, Japan.) This high-pressure logic system uses packless spool valves mounted on subbases that contain the interconnecting channels between elements. Each valve contains two windows in the form of the white circles on top through which spool position is visible.

3-5 THE IMPORTANCE OF TOTAL MACHINE COMMUNICATIONS

A fully automated industrial machine requires a feedback of information and a transmission of control signals, both to and from the logic circuitry. This two-way dialogue must be carefully planned so that selection of sensors and control actuators is given as much consideration as the logic circuitry components. Overall machine performance is dependent on the information received by the logic and the response to the commands.

The next chapter provides information on the peripheral equipment required to establish a complete dialogue with various parts of a machine. After this, Chapter 5 attempts a practical comparison between the various fluid logic systems described in this book.

CHAPTER 4

PERIPHERAL EQUIPMENT

4-1 THE NECESSITY OF PERIPHERAL EQUIPMENT

The term peripheral equipment encompasses all the devices apart from the logic hardware that is needed to create a complete industrial automation system. These devices include:

- Fluid supply treatment equipment.
- Sensors for feedback.
- Amplifiers to boost low-level signals.
- Interfaces—air-electric and electric-air.
- Time delays to delay starting and stopping of particular operations.
- Indicators and displays to keep the operator informed of the machine's operation and condition.
- Counters to keep a record of specific operations, to provide a control signal at a predetermined count, or to totalize the production rate.
- Readers and programmers to provide the logic with needed information.
- Fittings and tubing to facilitate construction and maximize operation.

4-2 TREATMENT OF SUPPLY FLUIDS FOR LOGIC DEVICES

The needs of various types of logic devices differs as to the requirements for filtration, water and oil removal, and pressure regulation. Figure 3-2 assists in defining the requirements above. We examine the individual needs of each system of logic devices defined in Chapter 3.

4-2-1 High-Pressure Movable-Part Devices

These devices have flow passage diameters equivalent to 0.1–0.2 in., and operating pressures within the 30 to 125-psig range.

Their requirements are similar to those of most standard pneumatic equipment:

- Filtration of coarse particles with a 40–50 μ sintered bronze element and the centrifical action of the water trap to remove all water condensation.
- Lubrication where needed, primarily for packed spool valves. Poppet-type logic devices do not require lubrication, and the development of self-lubricating materials eliminates the need for lubrication even in power valves and cylinders.

The standard configuration for applying these devices: filter, regulator, and lubricator. In complex circuits or where flow requirements are small, the use of standard oil fog-type lubricators may be required to reach all components. In this situation a "microfog" type lubricator is recommended because it emits a finer mist of oil which carries farther through the air lines before condensing back into oil droplets.

4-2-2 Low-Pressure Movable-part Devices

These logic devices have flow passage diameters in the range of 0.03 to 0.1 in. and operate on pressure between 10 and 30 psig. Good filtration and regulation is required to ensure proper operation. Originally these devices were designed to operate in conjunction with and from the same compressed air supply as pneumatic instrumentation.

To reduce contamination from distribution pipes, stainless steel or plastic pipes are usually used. In some applications the air is passed through dryers to lower the dew point to a specific temperature. At points of application the air is subjected to fine filtration to remove particles and oil. No lubricators are required in poppet valve logic systems. Standard regulators are adequate for most low-pressure logic systems.

Unfortunately, the quality of compressed air available to industries, in general, is not nearly as good as that required in instrumental applications. To ensure reliability and prolong the life of low-pressure logic systems employed in industry, the air treatment devices described should be incorporated into the control systems.

4-2-3 Jet Deflection—Nonmovable-Part Devices

Flow passage diameters in the range of 0.01 to 0.03 in. require special attention to the air quality. Clean, dry air is a prerequisite, and more accurate pressure regulation is required in the limited range between 2 and 10 psig.

Special filters known as "coalescing" types are rapidly becoming standard equipment on nonmovable-part systems to ensure the removal of fine particles and oil which line the passage walls of these devices, thus reducing their flow capabilities and disturbing the pressure balances required for their proper operation.

This problem could be minimized by higher operating pressures and larger flow passage. However, this would create an unacceptable new problem—higher air consumption due to the required continuous venting.

An alternative to the above is a low-pressure oilless air compressor used at each machine. Two types of compressors are available: turbine or diaphragm.

Since pressure requirements are small (usually 10 psig or less), less heat is generated than in standard compressors, therefore greatly reducing the water condensation problem.

4-2-4 Jet Destruction Nonmovable-Part Devices

The passage diameters are the smallest of all nonmovable-part devices and the operating pressures are the lowest, usually less than 0.50 psig. The pressure is reduced to 0.02 psig or less for the control function. These requirements make it clear that filtration and pressure regulation requirements are much more critical for this type of device. The normal air treatment requirements for other nonmovable-part devices is not adequate for use with jet destruction devices. To ensure air quality, double filtration is required (a coarse, then a fine filter), followed by an oil absorbtion filter, and then a low-pressure precision regulator.

As already mentioned, a low-pressure oilless air compressor is a better solution to the problem in this case.

4-3 SENSORS AND SIGNAL AMPLIFYING DEVICES

4-3-1 Introduction

The term input device includes sensors of all types: contacting and noncontacting, manual controls, temperature and pressure measuring devices, and such. They are required to produce digital outputs. Frequently signals from sensors are of an analog nature. In this case it is necessary that the sensor only switches at a predetermined threshold, or that the sensor feeds a device that is capable of converting analog to digital signals before the input is sent to the logic. A sensing device designed to respond to low-pressure input signals and convert them into high-pressure outputs is known as a "pressure amplifier." Usually pressure and flow are amplified simultaneously. We refer to this type of device as a power relay. Power relays are frequently employed at the outputs of low-level nonmovable-part fluid logic circuits to pilot power systems requiring high-pressure signals.

Devices are available that combine the sensing and power relay functions into a single device; this facilitates their use with high-pressure logic systems.

We simultaneously discuss sensors and signal amplifying devices, to point out their common and complementary operating principles.

4-3-2 Basic Operating Principles of Sensors and Power Relays

Figure 4-1 illustrates operating principles of movable- and nonmovable-part sensors.

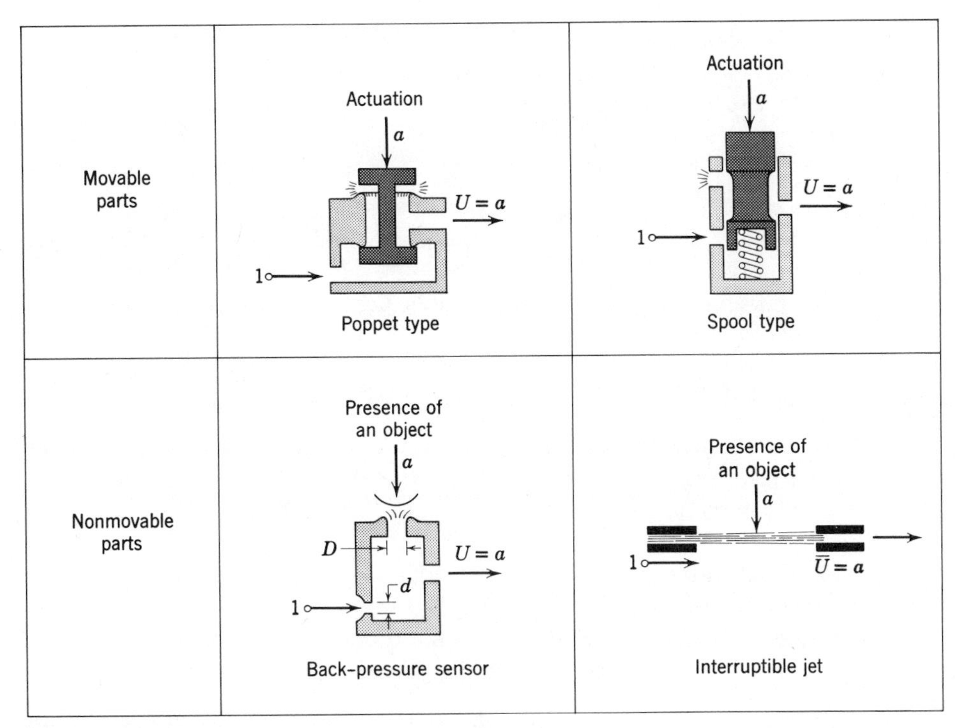

FIG. 4–1. Basic principles for sensors and power relays.

Movable-Part Sensors. The mechanical movement required to actuate the sensors is equivalent to the principles used in movable-part logic devices.

Nonmovable-Part Sensors. The actuation principles of the devices are quite similar to those used in the nonmovable-part logic devices.

- *Back-Pressure Sensors.* As illustrated in Fig. 4-1, a small chamber is fed with a compressed air supply through an orifice of diameter d. The exhaust port of diameter $D > d$ keeps the pressure from rising in the chamber. The object being sensed covers the exhaust hole, thereby increasing the pressure in the chamber until it flows through the output port. When the exhaust port is uncovered the pressure in the chamber drops and the output is reduced to zero.
- *Interruptible jet sensors.* A maintained supply jet is directed through a nozzle, across a gap to a receiving nozzle. When an object enters the gap and interrupts the air jet, the output of the sensor drops to zero. If the supply pressure is low enough the jet can be laminar as discussed in Chapter 3, for laminar/turbulent flow devices. The air jet usually employed in an interruptible jet is turbulent because the supply pressure is too high to maintain a laminar flow.

In practice, movable and nonmovable-part sensor principles are greatly modified by their manufacturers, as we explain.

4-3-3 Movable-Part Sensors and Power Relays

4-3-3-1 *Movable-Part Sensors*

There are two basic types of devices:

- The device known as "manually operated valve" (sensor) is usually mounted on a machine conveniently near the operator or on a control panel.
- The mechanically actuated movable-part sensor is known as a limit or trip valve.

Figure 4-2 illustrates a variety of standard manually operated valves (sensors).

If a three-way valve is used "normally closed," it provides an output signal when the operator actuates it. If a three-way valve is used "normally open", it provides no output when actuated and opens the output to exhaust to atmosphere.

A four-way valve provides the two inverse states of the normally open (NO) and normally closed (NC) three-way valves in each of its two positions.

Figure 4-2 also illustrates the basic manual operators used on three-and four-way valves. The two types of valves used with these devices follow:

- The spring return-type which changes state by manual actuation and returns to the first state when the actuator is released.
- The manual return type is maintained in the last position into which it was shifted by the operator. This last type may be referred to as a memory device.

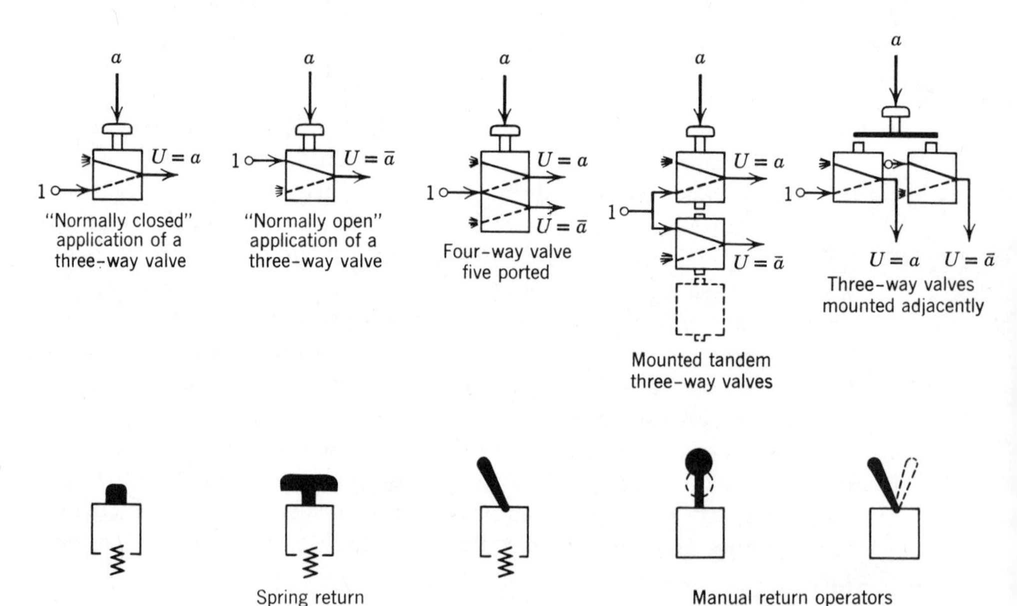

FIG. 4-2. Standard actuator styles and valve configurations for manually operated valves (sensors).

The variety of manual operators available is very similar to that of those available for electrical switches. In fact, some manufacturers design mounting frames that allow their valves to be attached directly to the back of electrical operators.

Figure 4-3 illustrates the basic types of mechanically actuated limit or trip valves (sensors).

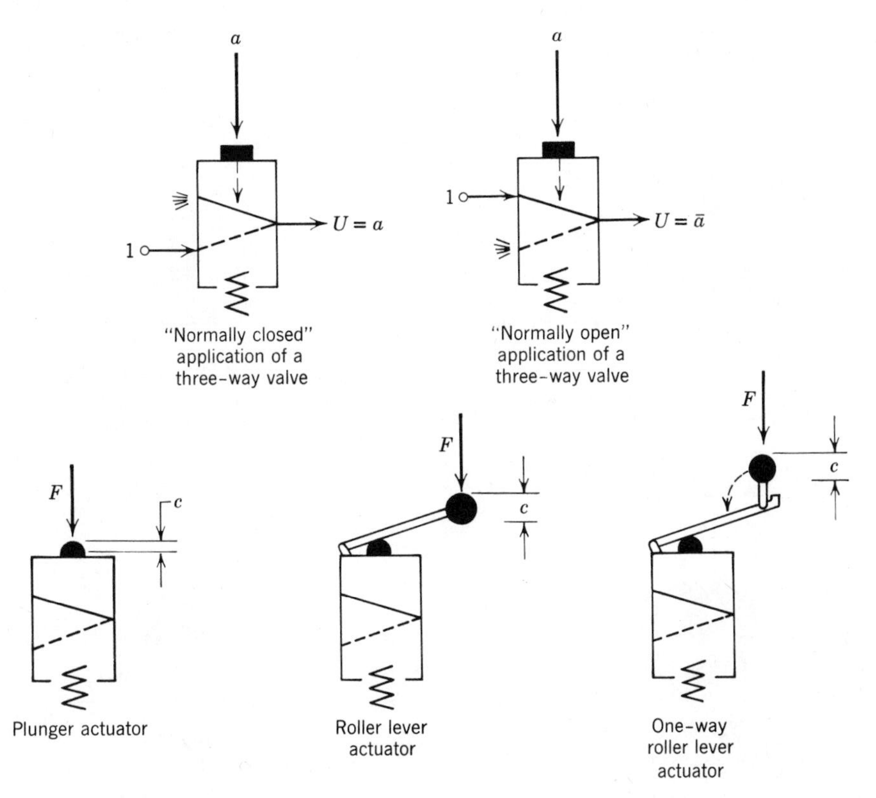

FIG. 4-3. Standard actuator styles of mechanically actuated movable-part valves (sensors).

In general, the size differential between three-way and four-way valves may exclude the use of four-way valves as sensors. The three-way valves may be applied in either the NO or NC mode.

Actuators for mechanically operated valves are available in a variety of types, and for all degrees of duty. The various manufactured brands vary in stroke length and force to actuate. The most common actuators are illustrated in Fig. 4-3.

Among the wide variety of mechanical actuators, the one-way roller lever is particularly valuable in fluid logic systems. The advantage of this style of actuator is discussed further in Chapters 7 and 8.

Figure 4-4 illustrates the primary valve designs used in conjunction with the manual and mechanical (sensors) operators just discussed.

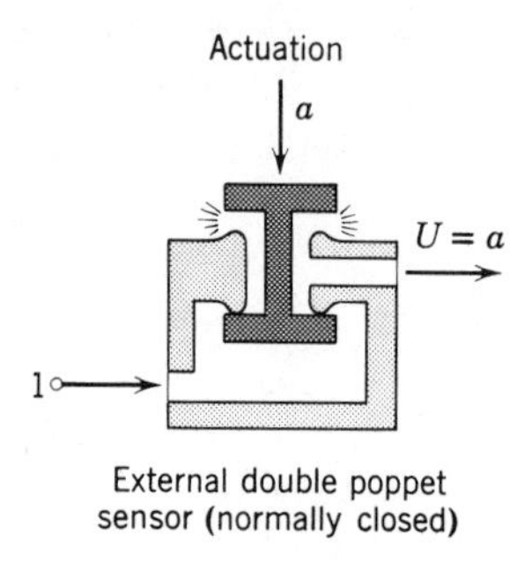

External double poppet sensor (normally closed)

Actuation

a

$U = \bar{a}$

1

Internal double poppet sensor (normally open)

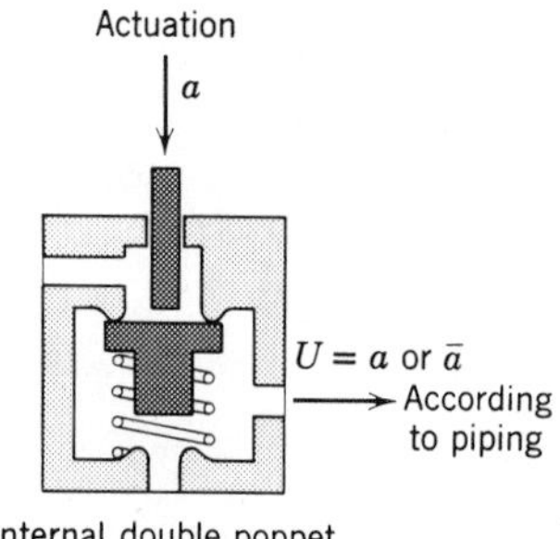

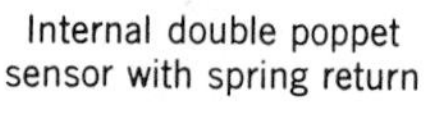

Internal double poppet sensor with spring return

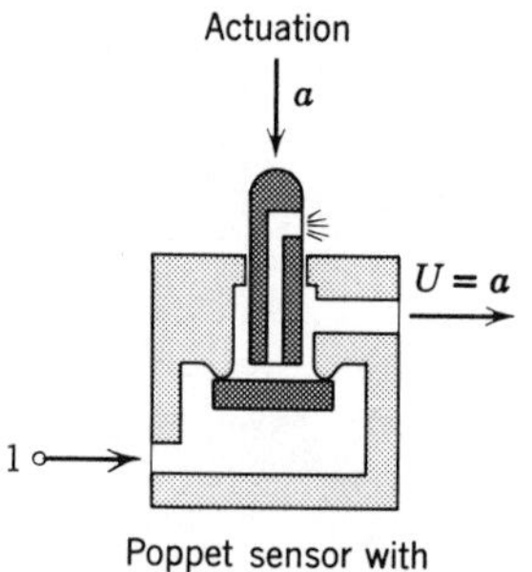

Poppet sensor with hollow actuator stem

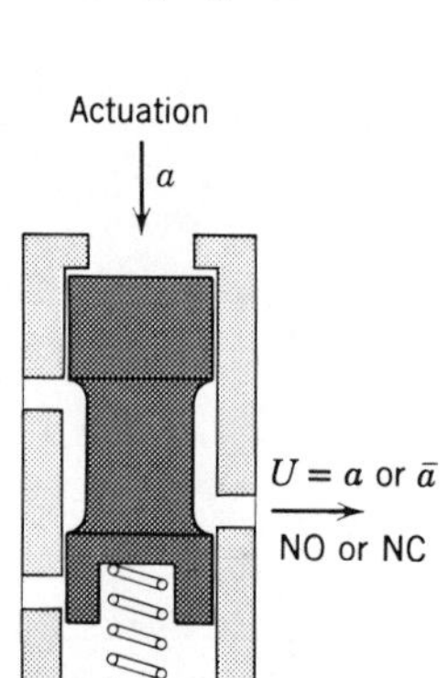

Spool type sensor

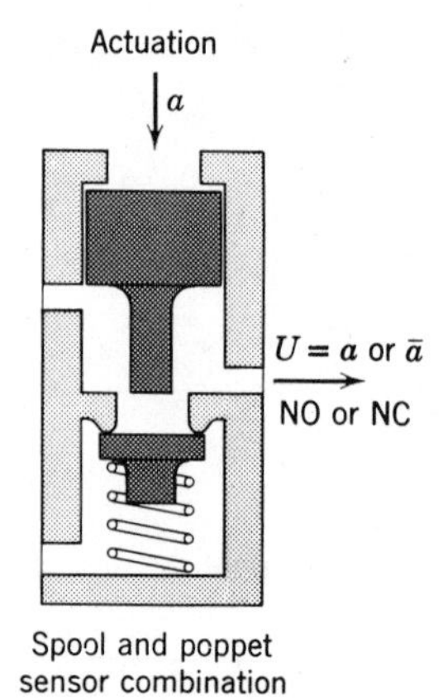

Spool and poppet sensor combination

FIG. 4–4. Valve designs for movable-part sensors.

- The external double poppet style can only be applied as an NC valve.
- The internal double poppet style can only be applied as an NO valve.
- The internal double poppet spring return style can be applied to an NO or NC.
- The poppet with the hollow actuator can be applied only as an NO valve. The advantage of this design is that the supply port flow is isolated from the exhaust port flow so that the exhaust port is closed before the supply port is opened when the valve is actuated.

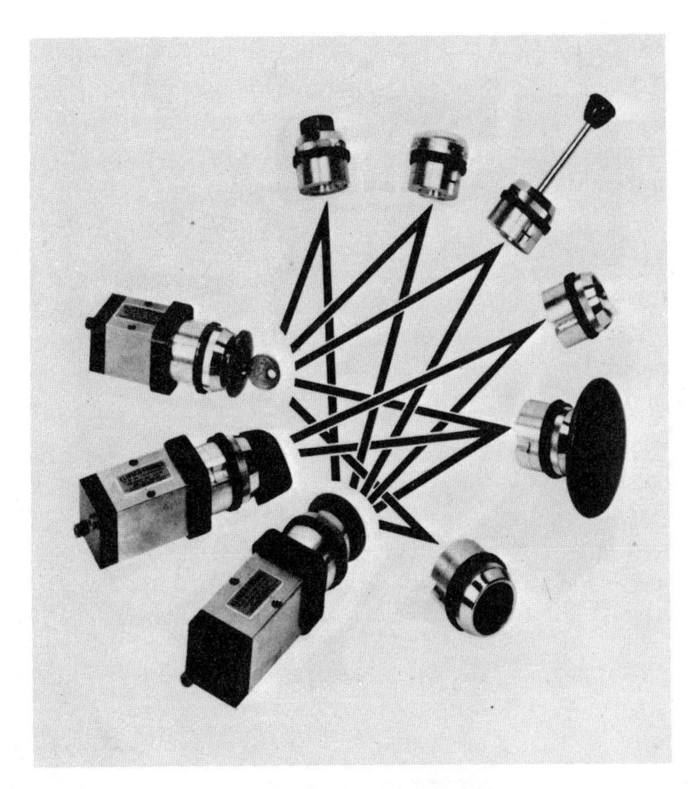

Movable-part sensors designed for manual operation. (Courtesy Climax-France, France.) This array of devices shows the versatility of actuator configurations designed to fit onto pneumatic valves.

Miniature Movable-part sensors. (Courtesy Crouzet, France and Miller Fluid Power, USA.) The group on the left is designed for mechanical actuation while those on the right are manually operated. Note on the right, under the panel that they can be stacked to operate each other. The basic element is the hybrid movable-part sensor amplifier described in Fig. 4-7.

Poppet-type limit valves require less travel to actuate than spool valves. However, the force required to actuate poppet valves increases proportionately with the increase in supply pressure. This undesirable effect is not a problem with spool valves, since spools are pressure balanced. The spool and poppet combination in Fig. 4-4 is a popular design.

4-3-3-2 *Movable-Part Amplifiers (Power Relays)*

Figure 4 - 5 illustrates some standard designs for movable-part *Amplifiers*—power relays. The basic design principles used in these valves have already been discussed under the section on NOT, AND, YES movable-part logic devices. Power relays can be considered basically as pressure relays.

- The YES function is created by a relay that provides a high-pressure output when a low-pressure is applied to the input.
- The NOT function is created by a relay which provides a high-pressure output when a low-pressure input is absent or removed.
- Some relays can be used in either way:
  YES = NC (normally closed) NOT = NO (normally open)

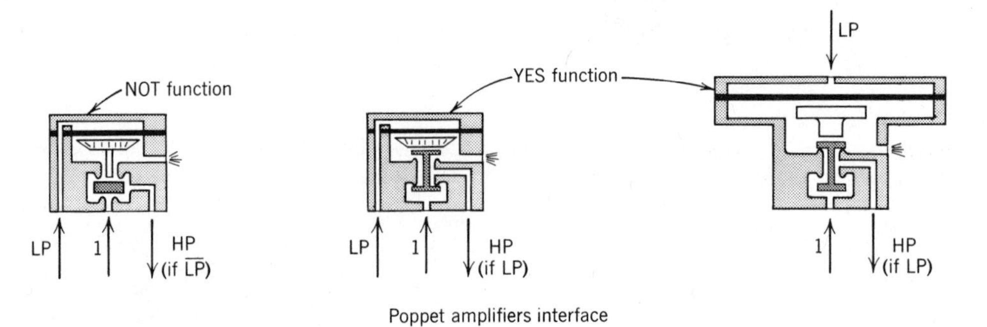

Poppet amplifiers interface

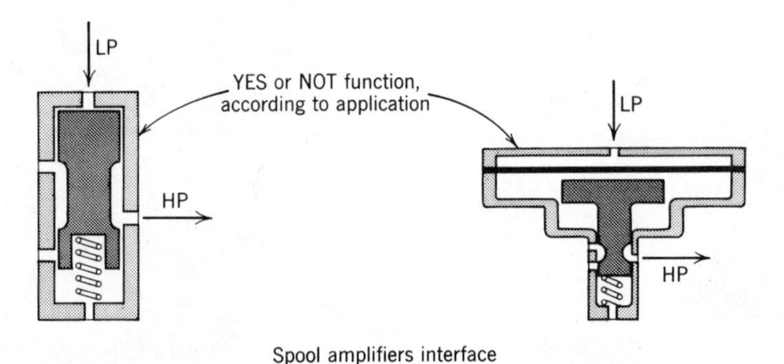

Spool amplifiers interface

FIG. 4–5. Standard designs of movable-parts amplifiers interface.

The amplification factor in poppet valves is determined by the ratio between the cross-sectional area of the diaphragm and the poppet.

In spool valves the amplification factor is dependent on the amount of friction of the spool packing and the force of the return spring.

Flow amplification is generally created by the differences in flow passage diameters of the input and the supply and output ports. We refer to these types of relays as power relays because of their ability to amplify both flow and pressure.

4-3-3-3 *Hybrid Movable-Part Sensor Amplifiers*

The manually and mechanically actuated sensors described above are not capable of being actuated by low levels of force. Hybrid movable-part sensors have been developed to meet this need. The devices are composed of sensors with built-in amplifiers. Since no generic name has been evolved for this type of device, we refer to them as "hybrid sensor amplifiers."

Figure 4-6 illustrates two common designs for this type of device. The first design employs a small poppet, which when actuated pilots a spool with a larger flow passage. The second design employs a small spool which when actuated pilots a poppet with a larger flow passage.

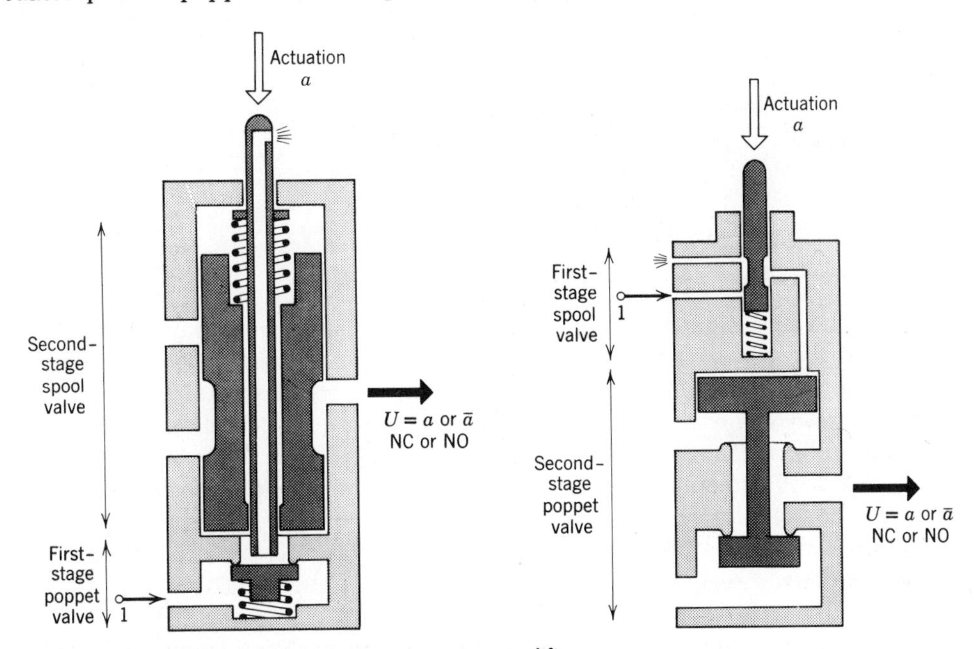

FIG. 4–6. Designs for hybrid movable-part sensor-amplifiers.

Figure 4-7 illustrates the actuated and deactuated states of these devices in a "stackable" design, similar to electrical switches. A small spool pilots, two diaphragms; each diaphragm acts on a single poppet; both poppets create a three-way switching action with a large flow passage. Since the bottom of the spool chamber is open, it provides an easy way to actuate an identical valve stacked directly under it. This stacking arrangement can be multiplied as required.

4-3-3-4 *Pressure Release Limit Valves*

Figure 4-8 illustrates the operation of a pressure switch employing a NOT logic device to sense the end of a cylinder rod stroke.

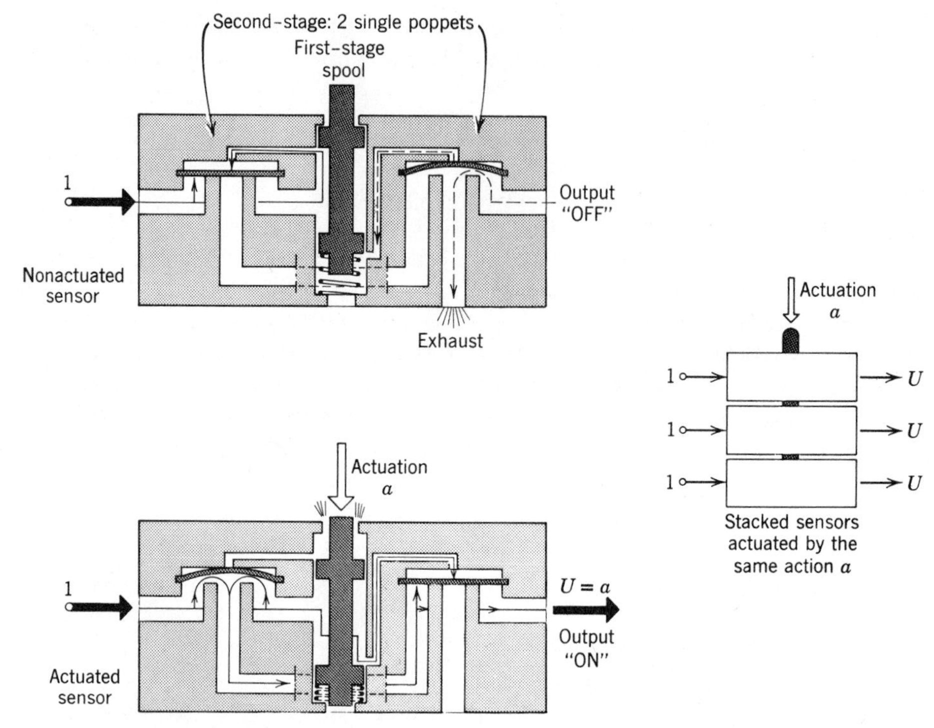

FIG. 4–7. Hybrid movable-part sensors-amplifier.

The curves on the graph represent the pressure changes on both sides of a cylinder piston as it moves through a stroke. The weight of the cylinder rod load and the friction factor determine the pressure differential across the cylinder piston while it is in motion. The pressure on the exhaust side of the piston drops to zero only at the end of the piston's stroke.

A NOT function device used as a pressure switch (sequence valve) receives the cylinder back-pressure (exhaust pressure) signal at its input port. The pressure to the input does not drop under the differential switching level until the piston reaches the end of its stroke: only then will the output switch "ON."

The pressure performs the same function as limit valves (trip valve) positioned for actuation at the end of a cylinder's stroke.

Another method of detecting the cylinder positions is to sense the "build-up" of actuating pressure in the cylinder (instead of the decrease of the exhaust pressure).

The use of pressure switches to create feedback signals for a sequential logic system is an alternative to limit (trip) valves, with their inherent mounting and actuation problems.

Although pressure switches are easier to apply, they have several disadvantages: if the cylinder seizes or is mechanically stalled before it reaches the end of its stroke, the pressure ahead of the piston will drop to zero, thereby actuating the pressure switch prematurely.

To minimize malfunctions when using pressure switches they are best applied where machine damage will not result from false pressure switch

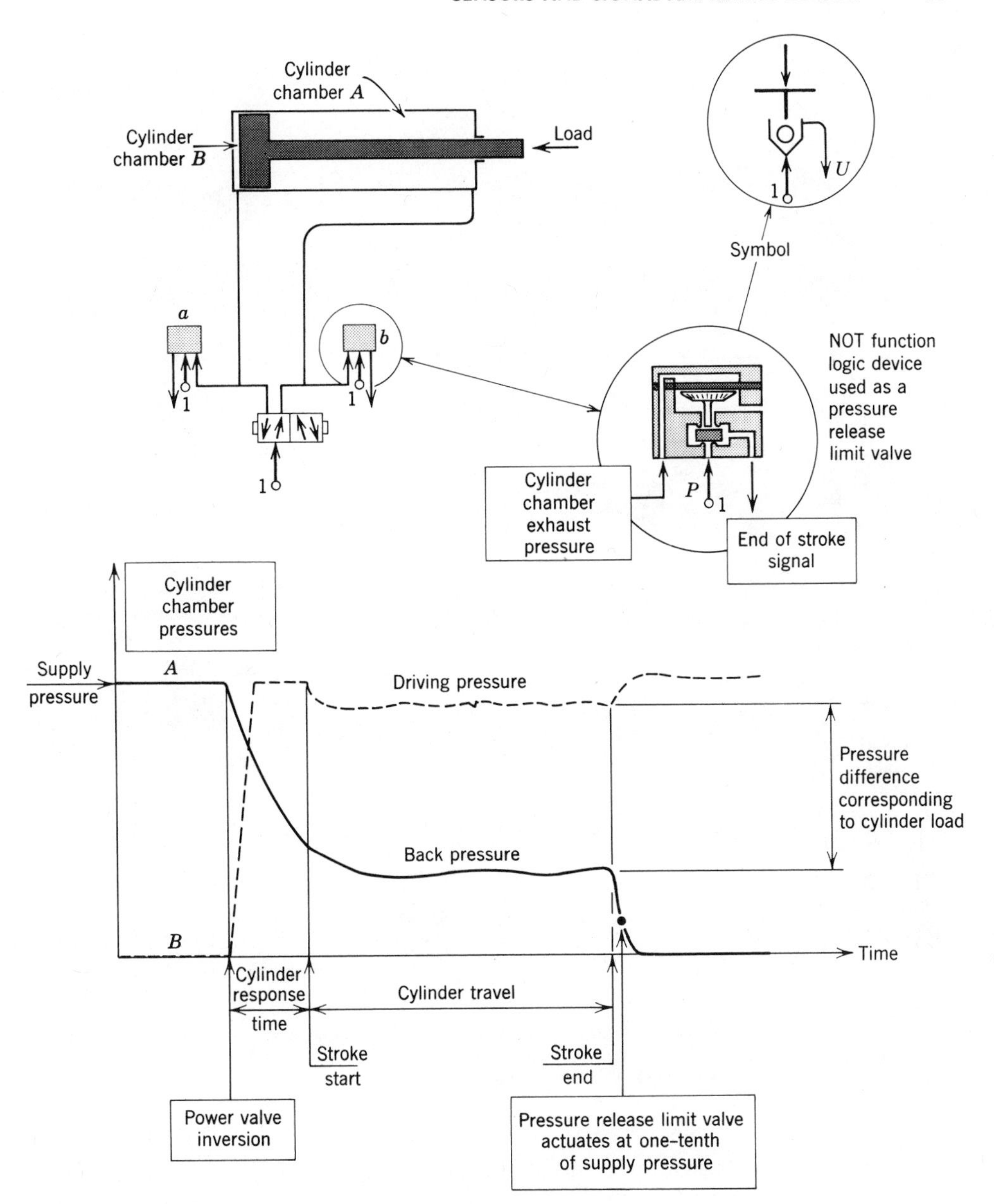

FIG. 4–8. Pressure release limit valve application.

output signals. Chapter 8 discusses typical applications for pressure release limit valves.

4-3-4 Nonmovable-Part Sensors

4-3-4-1 Back Pressure Sensors

We have discussed the simple principle required to create a back-pressure sensor—the blocking of the supply flow to atmosphere, and the resulting pressure build up used to switch "ON" a relay.

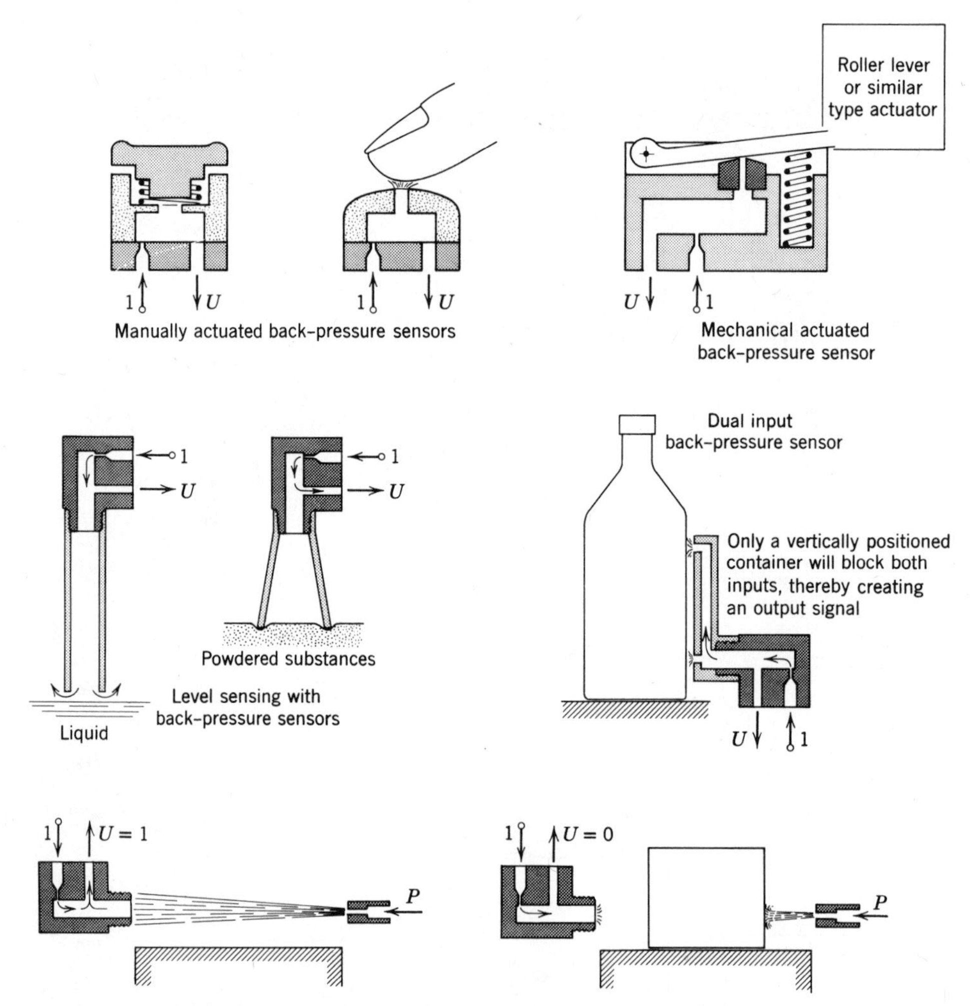

FIG. 4-9. Typical back-pressure sensors applications.

Figure 4-9 illustrates some of the typical applications of the back-pressure sensor principle.

- A manually actuated back-pressure sensor can be created by simply blocking the flow to atmosphere of a single supply line, with the touch of a finger.
- Mechanically actuated back-pressure sensors can be switched with short strokes and low force.
- Level sensing of liquids or powdered substances is easily accomplished with back-pressure sensors.
- The versatility of the back-pressure sensor is typified by the container position sensor, which employs two inputs. An output will occur only when both sensing ports have been blocked.

A back-pressure sensor can also be used as the receiver device in conjunction with an interrupting jet. This combination increases the sensing range of back-pressure sensors.

As illustrated in Fig. 4-9, simple back-pressure sensors can be used as limited range proximity sensors. They can switch "ON" if an object passes within 0.02 in. of the sensor's face. Partial blocking of the leak results in a rise of the output signal.

Advances in the design of back-pressure proximity sensors have greatly increased their sensing range. The basic operating principles of these devices are illustrated in Fig. 4-10.

The term "back-pressure proximity sensor" is usually shortened to just proximity sensor when the device employs a design other than straight orifice exhaust type and when the sensing range extends beyond a few thousandths of an inch.

The majority of proximity sensors employ an annular design for the

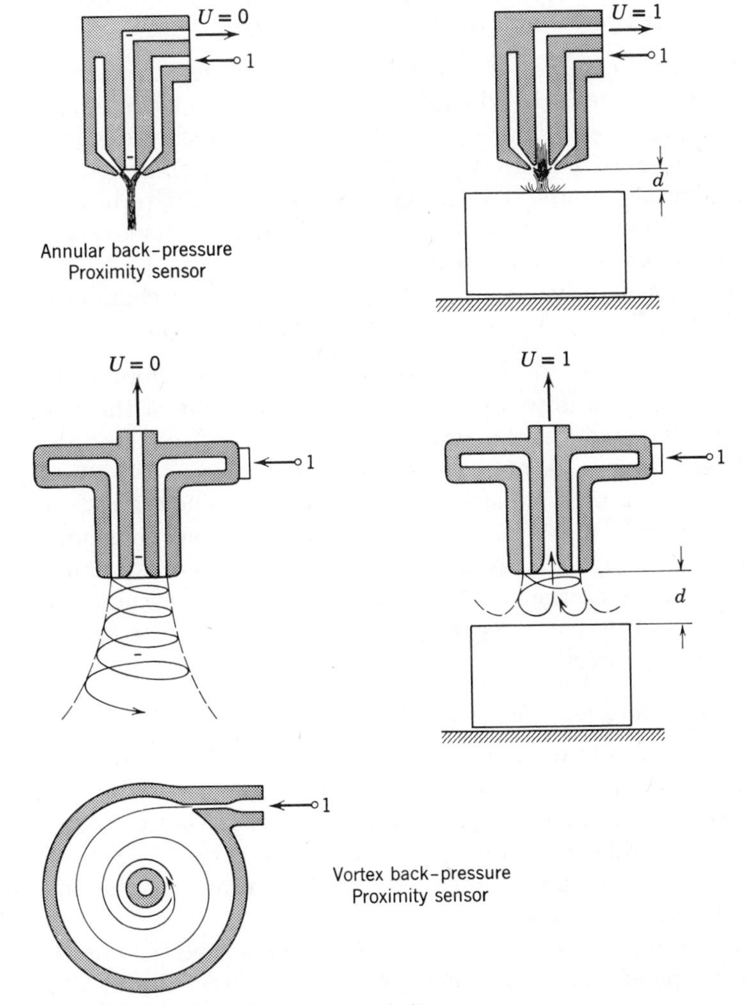

FIG. 4–10. Improved designs for back-pressure proximity sensors.

venting of the supply jet. The sensing port for the output signal is located in the center of the annulus. The venting of the supply creates a slightly negative pressure in the sensing port. When an object, preferably with a flat hard surface, is within the sensor range a small amount of the supply is reflected into the sensor through to the output port. The minimum range of most proximity sensors is 0.2 in. Most annular supply jets are focused into a cone shape to increase the sensitivity of the sensor.

The vortex proximity sensor vents its supply through an annular orifice in the form of a spiral. The spiraling action results in an extended sensing range. In practice this can project a usable jet up to about 0.5 in. The sensing characteristics are indentical to the cone shape type of proximity sensor.

Air guaging is well-established as an important sensing technique. It is frequently used to verify dimensions of precision parts in conjunction with analog-type control systems. Air guaging techniques have been used advantageously for sorting parts by size.

The air consumption of back-pressure-type sensors has been minimized by the following two popular methods.

Reduced Pressure. The operating pressure requirements of non-movable-part logic systems has facilitated the use of back-pressure sensors with low supply pressures. This type of application is highly desirable in that it eliminates the need for interface valves which inherently reduce the overall response time of sensors. The direct coupling of sensor to logic is ideal from several standpoints: lower sensor air consumption, faster overall system response, less components, and lower system cost.

The only negative aspect of low-pressure supply technique is the need for signal amplification at the output of the logic to the power devices.

Reduced Flow. As illustrated in Fig. 4-1, the diameter of the supply vent dimension D is large in contrast to the diameter of the supply port dimension d, which reduces the air consumption by orifice restriction.

When using back-pressure sensing principles on standard pneumatic pressure, the back-pressure sensor is a bleed valve. The diameter of d must be within the range of 0.004 to 0.010 in. if the rate of air consumption is to be acceptable. Air supply filtration has to be in accordance with these small orifices. When reducing the flow, and not the pressure, a contact-type sensing will provide full-pressure recovery, so that high-pressure logic devices can then be actuated directly.

4-3-4-2 *Interruptible Jet Sensors.*

The basic principle of interruptible jet sensors was described in Fig. 4-1. Variations and applications of these principles are illustrated in Fig. 4-11.

In contrast to movable-part sensors or back-pressure sensors, the operation of interruptible jet sensors is disturbed by aerosols such as dust or oil. These contaminents can be ingested by the sensor and thereby restrict or completely clog the output port.

To minimize this problem, opposed interruptible jets have been developed. This type of sensor consists of two opposing jets of unequal supply pressures colliding in a plane at the input of standard back-pressure sensor.

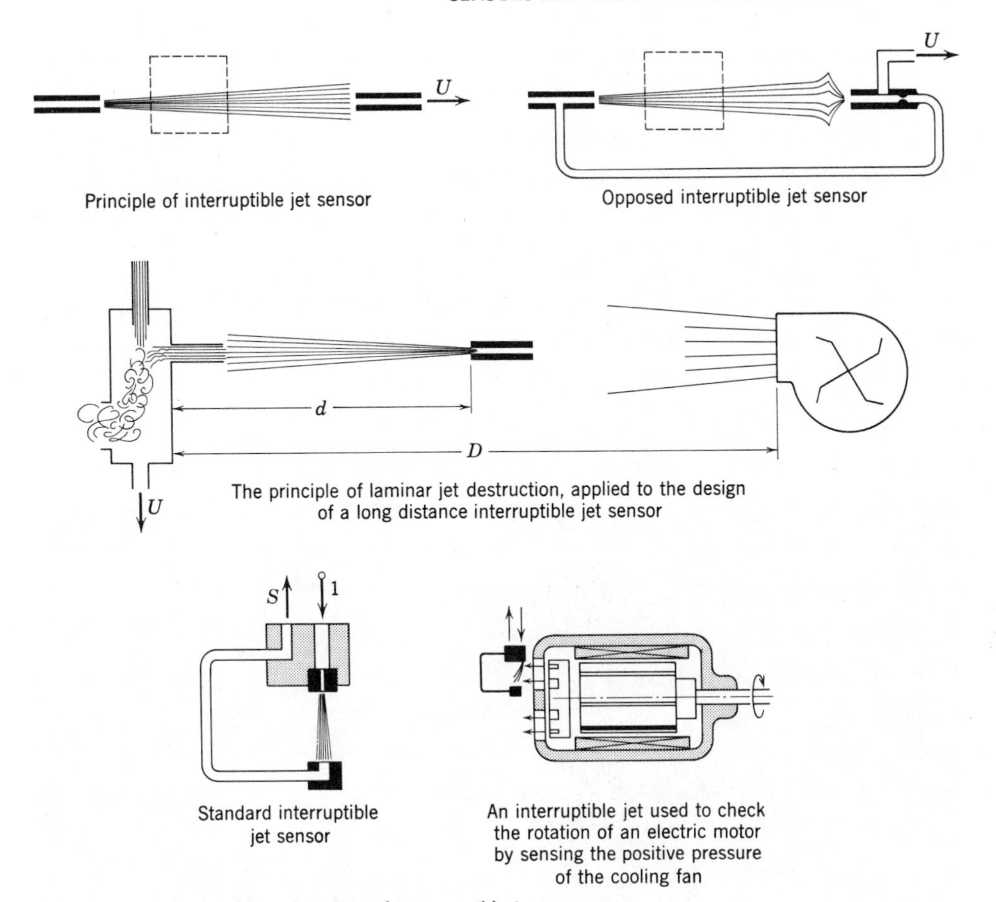

FIG. 4–11. Design and applications of interruptible jet sensors.

This blocks the output of the sensor, thereby back pressuring the supply into the output port *U*. When an object interrupts the higher pressure main jet, it removes the impinging air jet thereby allowing the supply to vent and the output of the sensor to drop to zero.

This type of sensor is known as an opposed jets sensor. Its main feature is the self-purging action, which minimizes the tendency to ingest contaminents.

The use of a laminar jet in conjunction with an interruptible jet illustrated in Fig. 4-11 provides a means of creating a long distance sensor. The air consumption of a high-pressure jet is acceptable up to about 3 ft. After that, a blower is recommended as the air supply source. This combination is good up to about 15 ft, with acceptable power consumption.

The versatiliy of the interruptible jet sensor is illustrated by the electric motor rotation sensor in Fig. 4-11. The sensor jet is interrupted by the positive pressure of the motor cooling fan.

The potential applications for air sensors is more readily appreciated by comparing them with their basic equivalents in electrical sensors:

- The interruptible jet sensors are very similar to the photo cells and allow long-range sensing.

- The back-pressure sensors are similar to magnetic or capacitor-type electrical proximity sensors.

4-3-4-3 *Ultrasonic or Acoustic Sensors.*

The disruption of laminar jet streams by sound wave was mentioned in Chapter 3. Fluidic technology has utilized the basic principles of acoustics to create an ultrasonic sensor.

As illustrated in Fig. 4-12, an emitter (whistle) produces an inaudible tone (in the range of 50 khz) The sound wave is detected by a receiver which employs the laminar/turbulent flow principle. The laminar jet is destroyed by the tone, acting as an input signal. The output of the receiver is normally "off" when an object is not blocking the path between it and the transmitter. The laminar jet flow to the output of the receiver is reestablished when an object blocks the passage of the sound wave to the receiver.

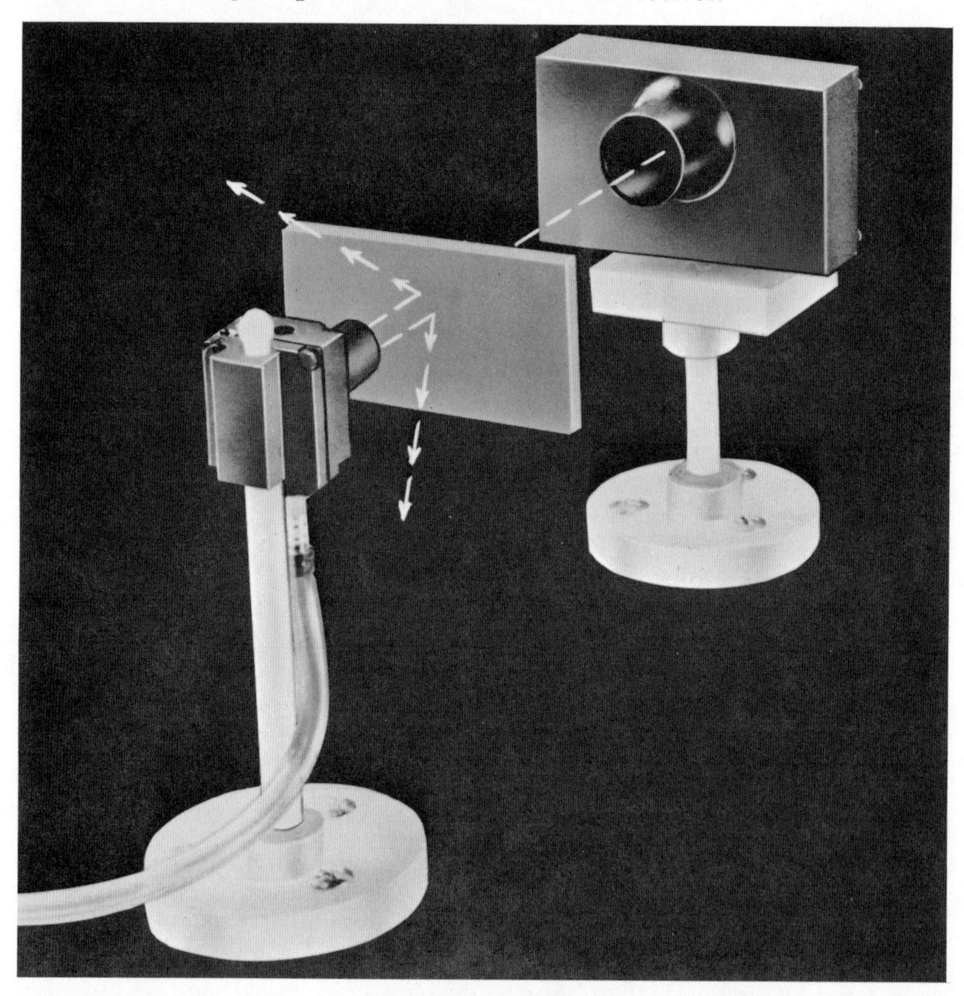

Ultrasonic fluidic sensors. (Courtesy ASCO Fluidics, USA, formerly Pitney Bowes.) Known as the "fluidic ear", this sensor operates according to the basic principles described in Fig. 4-12. Both transmitter and receiver are shown in this view.

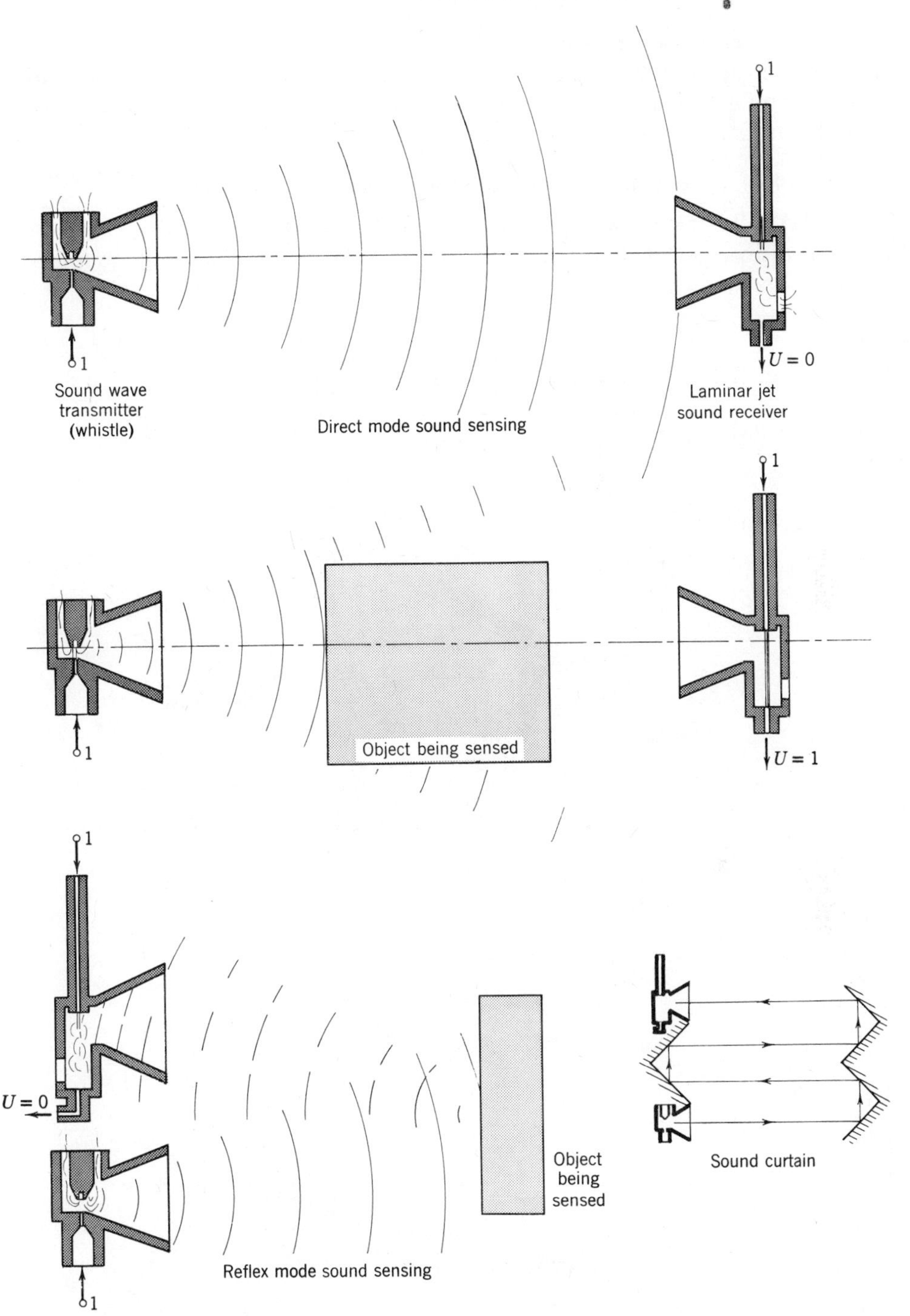

FIG. 4–12. Ultrasonic fluidic sensors.

As illustrated in Fig. 4-12 the properties of sound waves extends the usefullness of the ultrasonic-type sensor: the reflex method of sensing can detect objects on a conveyor line or a sound curtain can detect the presence of a person entering a dangerous area.

Sonic sensors currently available have sensing ranges up to 5 ft in the direct mode, up to 1 ft in the reflex mode, and up to 20 ft when used with parabolic reflectors to extend the range for special applications such as "sound curtains."

4-3-4-4 Special Applications for Non-movable-Part Sensors

Figure 4-13 illustrates two types of jet sensors designed to operate in conjunction with a gauge dial. The first one uses an interruptible jet mounted in a fixed position.

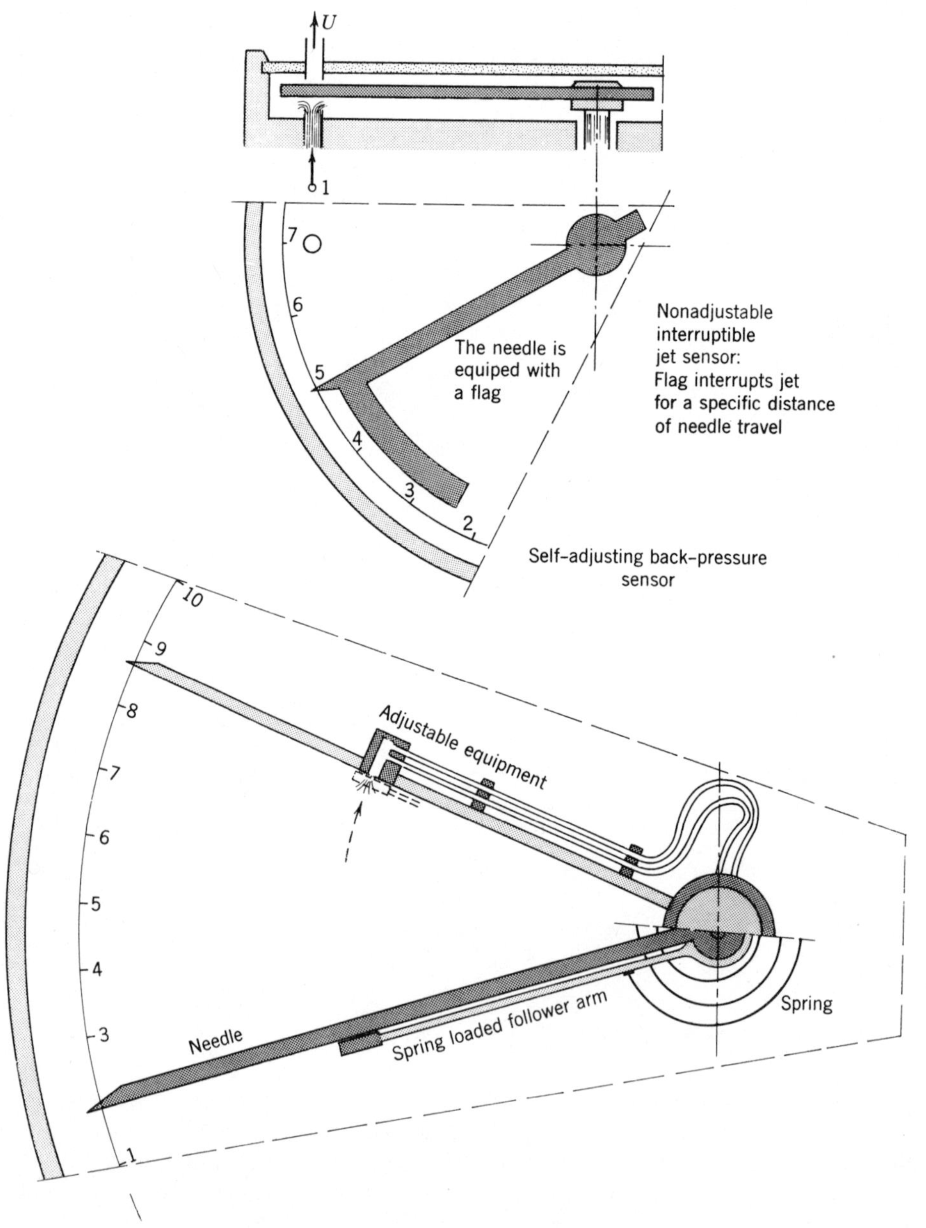

FIG. 4–13. Jet sensors applied to gauge dials.

The second device employs a back-pressure sensor which can be adjusted on the scale to sense the position of the needle.

The high sensitivity of nonmovable-part sensors is invaluable in gauge dial sensing applications. Jet sensors provide a means of directly linking all gauge dial readouts with fluid logic inputs. These sensing methods can be applied to thermometers, manometers, speedometers, accelerometers, weight scales, and such.

Punched card reading (illustrated in Fig. 4-14) is simplified by means of either of two nonmovable-part jet sensors. The reading of punched cards or tapes with jet sensors provides a direct means of feeding complex machine programs into fluid logic systems.

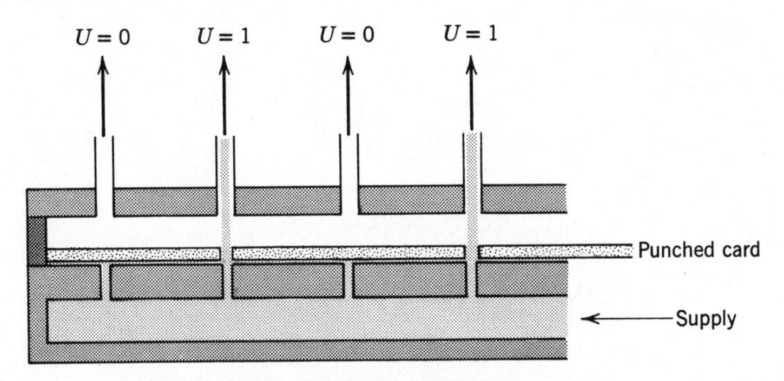

Punched card reading with interruptible jets

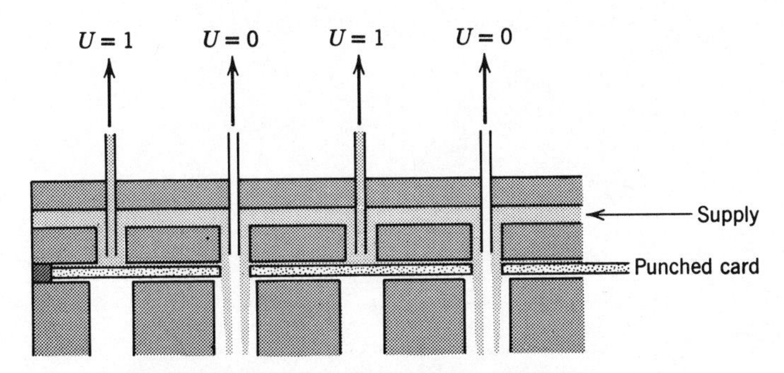

Punched card reading with back-pressure sensors

FIG. 4–14. Punched card reading with fluid logic devices.

4-3-5 Hybrid Power Relays

Relays or amplifiers employing only nonmovable-part techniques have been discussed in Chapter 3 as logic relays. Since their ability to amplify is very limited, they cannot be truly termed power relays.

Complete fluid logic systems employing only nonmovable-part sensors and logic must use high-pressure power relays to amplify the pressure and flow up to levels required to operate industrial fluid power equipment.

A typical power relay is illustrated in cross-section in Fig. 4-15. It can increase a 0.02-psig input signal into 100-psig output signal by combining two stages of amplification:

- The first stage is a back-pressure diaphragm amplifier. The diameter of orifice *d* defines the required venting flow (continuous air consumption on when input not actuated). The upper diaphragm is depressed by a low-pressure input signal. This action blocks the flow from orifice *D*.
- The second-stage diaphragm poppet assembly is actuated by the back pressure from the blocked orifice *D*. The double external poppet shifts downward opening the high-pressure supply to the output port.

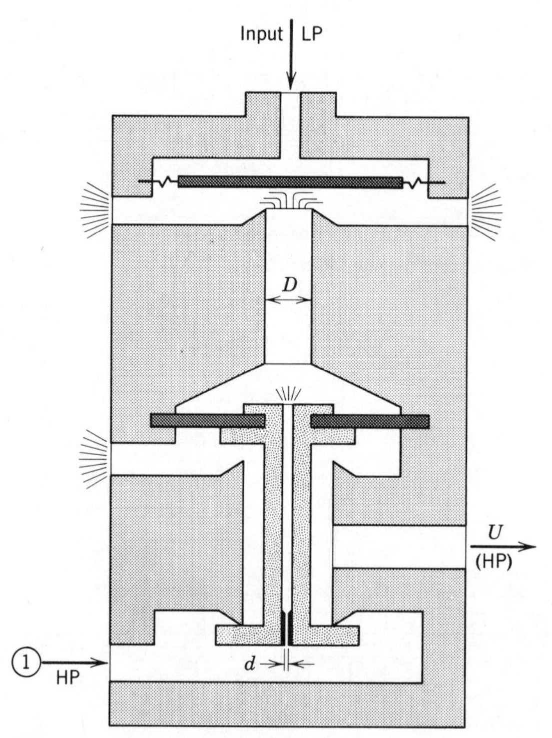

FIG. 4–15. Hybrid power relay.

The hybrid relay illustrated in Fig. 4-16 combines movable-part and nonmovable-part venturi principles. It could be employed in connection with an interruptible jet or back-pressure sensor. It operates from a high-pressure supply (60 to 90 psig) and uses the supply to operate the sensor and to amplify the sensor output.

The first stage is actuated by sensor output signals as low as 0.02 psig.

The second stage uses two double poppets to create a four-way valve. The unique feature of this relay is the direct connection between the two am-

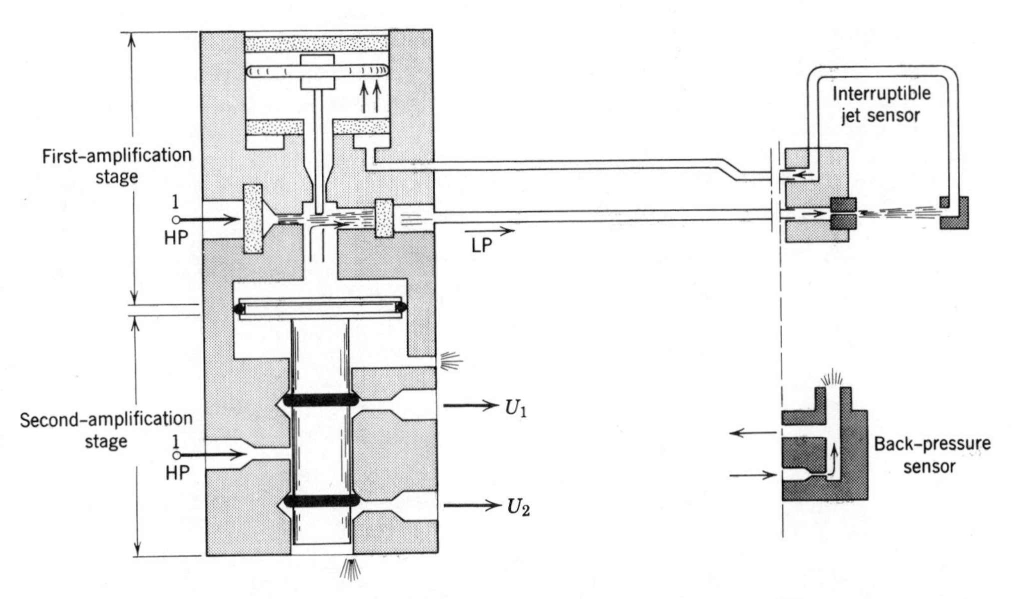

FIG. 4–16. A special power relay provides the low-pressure supply for nonmovable-part sensors and the amplification of their output signals.

plification stages. The supply is restricted as it enters the valve. The supply passes through an area with a large cross-sectional diameter; this sudden expansion creates a venturi effect which causes a negative pressure level (8.5 psig vacuum). This partial vacuum pulls up on the second stage poppet, thereby shifting the valve. This same supply air, greatly reduced in pressure, passes out of the opposite port of the relay to supply the nonmovable-part sensor.

When the output of the sensor drops to zero (due to an object interrupting the jet), the suspended piston and rod assembly drop by gravity. The rod drops into the path of the supply thereby destroying the venturi effect. This action converts the 8.5-psig vacuum into an 8.5-psig positive pressure (due to the resistance to flow). This pressure acts on the second stage poppet assembly to shift the valve outputs.

4-3-6 Hybrid Movable- and Nonmovable-Part Sensor Amplifiers

Since nonmovable-part sensors can only create low pressure output signals, it is advantageous to combine them with movable-part amplifiers into a single device that we call a sensor amplifier.

Figure 4-17 illustrates three examples of this type of device.

A Back-Pressure Sensor-Spool Valve Amplifier Combination. The first stage is a mechanically actuated back-pressure sensor; the second stage consists of a differential pressure piloted spool valve. In the nonactuated position, the supply flows through the spool to shift and hold the spool to the left.

The output is then connected to the exhaust port. In the actuated position, the supply shifts the spool to the right. The supply can flow directly to the output. The short stroke length of the mechanical actuator is sufficient to create a high-pressure output.

An Interruptible Jet Sensor and Diaphragm Amplifier Combination. The working principles are clearly illustrated in Fig. 4-17.

A Vortex Sensor and Back-Pressure Diaphragm Amplifier Combination. The principles are clearly illustrated in Fig. 4-17.

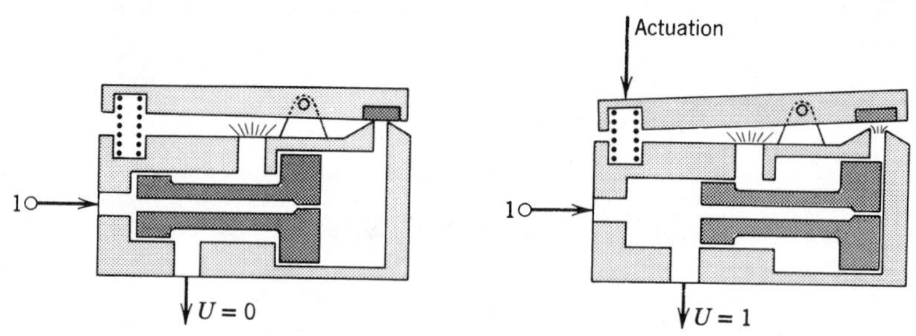

Combination back-pressure + spool valve sensor amplifier

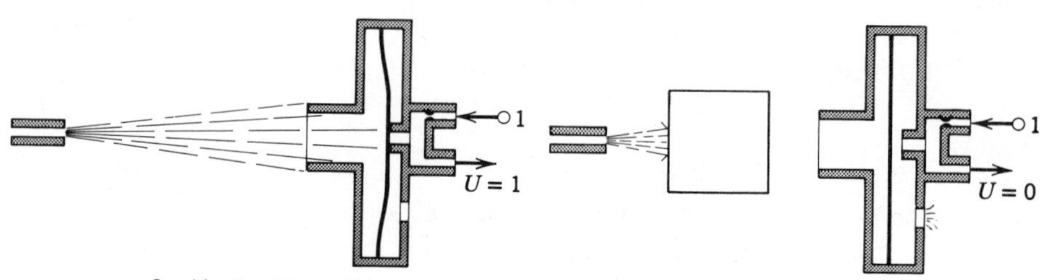

Combination interruptible jet + back-pressure diaphragm sensor amplifier

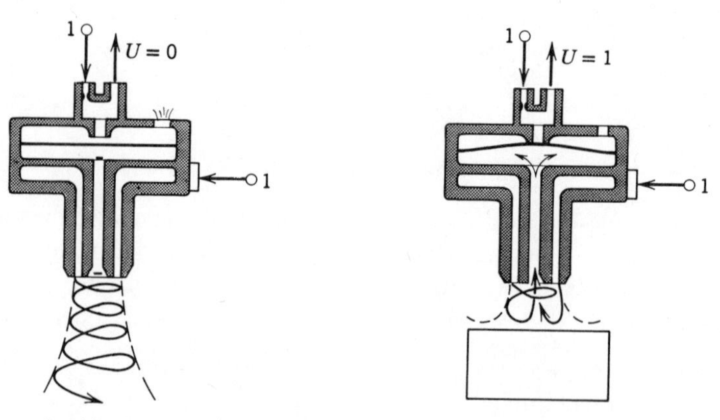

Combination vortex + back-pressure diaphragm sensor amplifier

FIG. 4-17. Hybrid sensor amplifiers.

Low-pressure fluidic tape programming system. (Courtesy Festo, Federal Republic of Germany.) This view illustrates the concept of perforated tape programming to control an industrial sequence. The array of visual indicators on the top panel identify progression through the program format.

4-3-7 The Fluid Logic Type Schmitt Trigger—An Adjustable Analog to Digital Signal Converter

A Schmitt trigger created with fluidic devices provides adjustable triggering settings and sensitivity to low-level input signals. As mentioned earlier, a practical application of the Schmitt trigger is in conjunction with back-pressure sensors for dimensional verification of components, for the purpose of sorting according to size. The output of the back-pressure sensors are adjusted to switch the Schmitt trigger at specific settings, which correspond to the dimensions of the component.

Figure 4-18 illustrates the operating principles of a fluidic Schmitt trigger created with three jet deflector analog amplifiers and one monostable digital wall attachment amplifier.

The operating principle of a jet deflection analog amplifier is illustrated

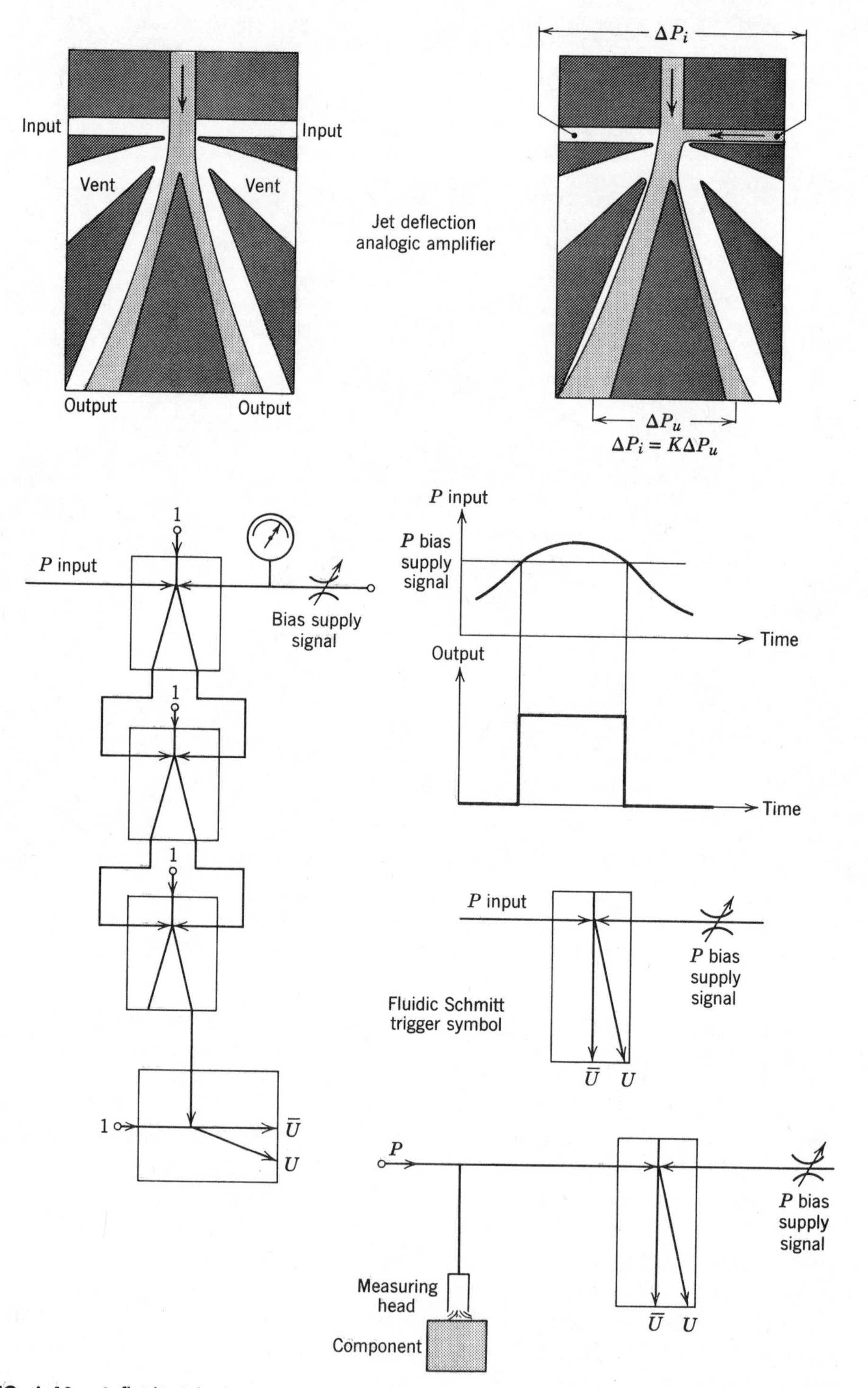

FIG. 4–18. A fluidic Schmitt trigger, consisting of three analog amplifiers and a monostable digital amplifier.

in Fig. 4-18. The outout pressure differential ΔPU is an amplified and proportional version of input ΔPi. .

$\Delta PU = K\Delta Pi$ (K varies within 5 to 10, according to type of amplifier)

The triggering set-point is determined by the adjustment of the bias supply signal. The analog input opposes the bias pressure; the difference between these two signals is amplified by the three amplifiers. The states 1 or 0 are available at the outputs U and $\overline{U}$.
A differential pressure as small as 0.002 psig can be amplified. This sensitivity can be utilized with air gauging techniques for highly accurate dimensional verifications. The four amplifiers used in the Schmitt trigger can be engraved into a single logic device. The supply, input, and output ports are indicated by the function symbol.

Chapter 8 describes two typical applications for fluidic Schmitt triggers in fluid logic systems.

4-4 INTERFACING WITH ELECTRICAL DEVICES

In industrial automation it is frequently necessary to interface from electrical signals to pneumatic inputs or from pneumatic outputs to electrical switches. Two common types of interface devices are required to perform these functions:

- The solenoid valve.
- The air to electrical switch.

4-4-1 Solenoid Valves

A solenoid valve converts an electrical signal into fluid flow, and Figure 4-19 illustrates the basic operating principle where a double internal poppet controls flow from a supply port to the output port. A spring holds the poppet

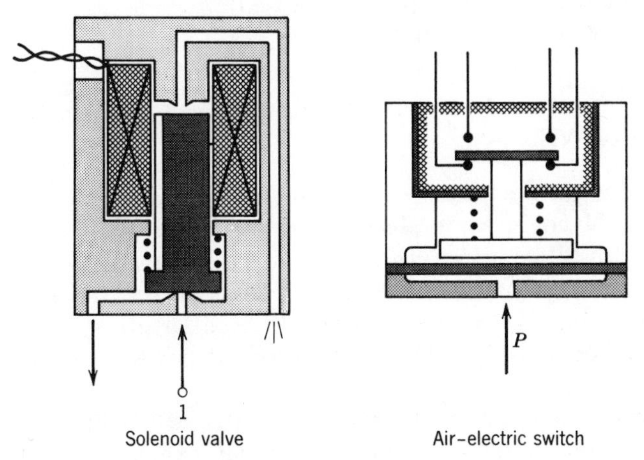

FIG. 4–19. Fluid-electric interface components.

down, internally blocking the entrance of the supply. The upper half of a steel poppet actuator which is attached to the poppet is surrounded by a coil of wire. When electricity is passed through the coil, a magnetic field is created which attracts the steel actuator, thereby pulling it upward. This lifting of the actuator-poppet assembly compresses the spring and opens the passage between the supply and the output.

Solenoid valves have been used extensively in electromechanical control systems. Their sensitivity and reliability have been improved through the years.

4-4-2 Air to Electric Switches

Their function is to close-or open-electrical contacts when an air input signal is received. Figure 4-19 illustrates their operating principles. The majority of them are designed to actuate with a given range of pressures.

A particular group of these devices have an adjustment for setting the pressure point at which the contacts will actuate.

4-5 TIMING DEVICES

The timing function is frequently an important circuit requirement in industrial automation systems. Time delays ranging from fractions of a second to several minutes may be necessary to supplement the operation of the logic control system

Examples of timing include:

- Pulses to operate logic or power valves.
- Several seconds for dwell time on a drill.
- Variable delay times for the cooling cycle on plastic molding machines.

Two basic timing principles are used:

- The adjustable orifice and volume chamber used in conjunction with a "snap-action" pneumatic relay.
- Motor or spring-driven timing devices.

4-5-1 Timers Employing the Principle of an Adjustable Orifice Metering into a Fixed Volume

Figure 4-20 illustrates a principle similar to the *R-C* function (resistance capacitance) employed in some electrical timers. The input signal is metered through an adjustable orifice that is connected directly to a fixed volume chamber. The output of the chamber is connected to the pilot section of a "snap-action" valve. When the required differential pressure level is reached the valve switches its output "ON" or "OFF" depending on the timing function required.

The time delay period is controlled by the adjustable orifice. When the input signal is removed, the volume chamber exhausts through the check valve. The timer is now reset, ready for the next cycle.

This type of timer is sensitive to fluctuation in input or supply pressure levels. An example of this situation would be when the supply pressure increases, the volume chamber will be filled more rapidly. Compensation for this variation factor is made by employing a "snap-action" relay which operates solely on the principle of differential pressures.

The standard tappered needle valve design employed in most variable orifice restrictors is not recommended for use in pneumatic timers because of its propensity to clog. To minimize this potential problem, "Vee" groove needles are employed as orifice restrictors.

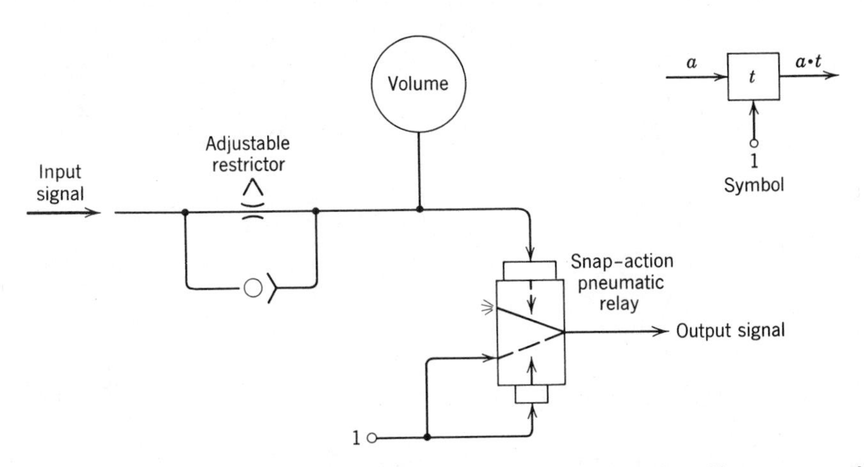

FIG. 4-20. Operating principle of timers employing the combination: adjustable restrictor, volume, and snap-action relay.

Figure 4-20 illustrates a timer employing air from an external supply source. The potential contamination problem of this system has been overcome by a timing device which operates on a slightly different principle. A volume of air contained inside the device is recycled each time the device is operated. This eliminates the possibility of clogging the adjustable orifice with contaminants. This same principle is quite commonly employed in electrical timers. As a result this timer has been improperly called a pneumatic timer.

Pneumatic timers, employing the principle of an adjustable orifice filling a fixed volume, are the most common type of pneumatic timers. Their simple design has proven itself dependable, relatively accurate (2 to 5%), and compact and easy to use in industrial applications. One defficiency of this device is the inherent impossibility of equiping it with an accurate scale for time indications.

4-5-2 Motor or Spring Driven Timing Devices

Figure 4-21 illustrates in part an escapment-type timer and a motor-driven-type timer. The escapment type timer works as follows: the input signal actuates a single cylinder that releases the escapment mechanism which rotates the timing disc. A bleed valve is mounted on an adjustale arm which is positioned to predetermine the time delay period. When the stop on the timing disc contacts the bleed valve, the three-way valve switches "ON" the output signal of the timer. The motor-driven timer utilizes the same basic principle as shown on Fig. 4-21. The constant speed motor can be driven either electrically or with fluid power.

The pneumatic timer employing the mechanical escapement principle is much more accurate than the adjustable orifice, fixed volume type. The accuracy of this type of device is usually better than 1% of the dial setting. In fact they are usually as accurate as electrical synchronous motor timers. The motor-driven timer utilizes the same basic principle, as shown on Fig. 4-21.

Cam programmers or card readers employing constant speed motors for actuation can be considered "multitimer" systems, as illustrated in Fig. 4-21. The constant speed motor can be driven either electrically or with fluid power.

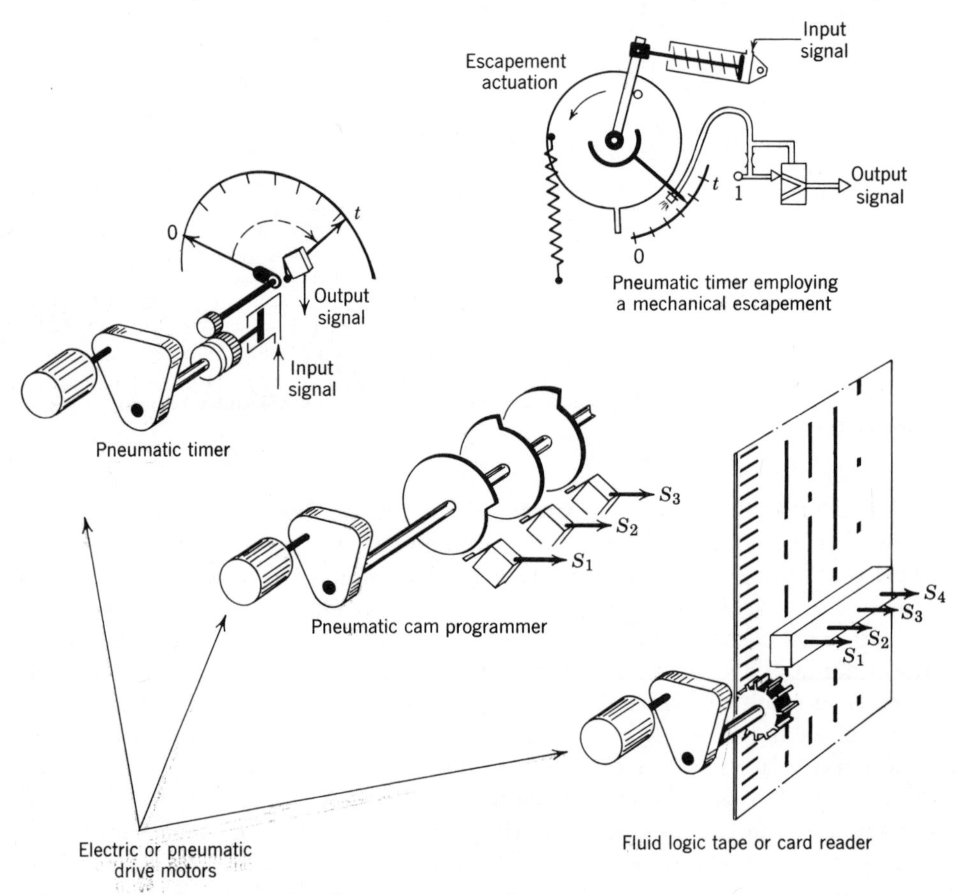

FIG. 4-21. Operating principles of various timers and programers.

In most applications, the synchronous electrical motor can be employed, its continous rotation eliminates the need for electrical relays to be switched for each step in the cycle.

In applications where electrical devices present a hazard, a pneumatic turbine, in conjunction with a watts speed regulator applying the principle of centrifical weights can be used to provide a constant speed drive.

4-6 PROGRAMMING DEVICES

Industrial automation systems sometimes require several different control programs on the same machine. An example of this is a machine that can perform various machining operations on a variety of products, thereby requiring a different control program for each product.

Programming devices are designed to faciliate the construction of the types of machines above, which are able to switch from one program to another. They are able to solve the program reading problems separately or simultaneously and the requirements to switch easily from one program to another.

In our discussion of sensing devices we have seen the simple principle of the punched card reader. Each card contains the required information for a complete program.

Low-pressure numeric display. (Courtesy ASCO Fluidics, U.S.A., formerly Pitney Bowes.) This view illustrates a form of low-pressure numeric readout devices. Commonly used with completely pneumatic counters, they operate directly from the flow mode elements pictured below them. Alphanumeric symbols are created from the standard seven bar code format and visibility is reflected from the ends of pistons that otherwise withdraw into a matrix of 35 cylinders when no readout is present.

Simple programs can be handled by standard logic circuitry alone. Manually operated valves can be employed to switch from one program to another. One alternative is to employ a manually operated pneumatic commutator which can be constructed easily with cams and sensors as illustrated in Fig. 4-22.

Continuously rotating programers do not use feedback signals. This type of device can be implemented by cams or punched cards or tape programs, as illustrated in Fig. 4-21. Here actuation is accomplished by a continously rotating drive motor. This is a simple technique; however, it does not result in very reliable automation, since the reading of the program continues even if one or more of the required actions is not completed.

Sequenced Program Retrieval. In each step of the program, specific actions are initiated. When all actions are completed, feedback signals from the sensors switch the programmer to the next step in the sequence. This method of program retrieval is illustrated in fig. 4-22. The feedback signal actuates a small single acting cylinder that steps the outputs a step at a time,

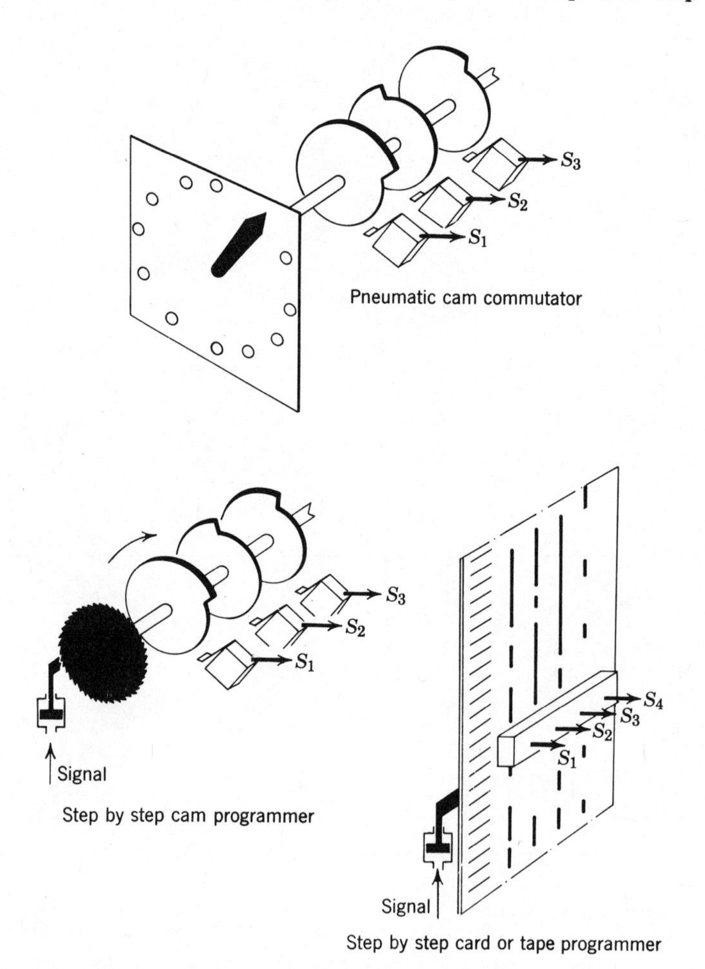

FIG. 4-22. Programming devices.

either on a cam shaft (cam programmer) or a punched card which is illustrated and becomes a punched tape (a tape reader).

Sequencers. The sequencer is a special type of programmer. The primary difference between the two is that sequencers provide only one output at a time, during each step in the sequence. The sequencer receives stepping signals in the same manner as the programmer. Each input pulse switches on an output and switches off the previous output; this results in only one output being "ON" during each step in the sequence of events.

Figure 4-23 illustrates a special revolving type sequencer. A small cylinder A actuates two air supplies simultaneously by a revolving distributor system: One of the distributors' output supplies a high-pressure power or piloting signal. The other distributor output provides the back-pressure sensor supply and feedback signal.

For each step in the sequence an output signal is provided to pilot a power valve or operate directly a small air cylinder. When the back-pressure sensor is actuated (only one is active during each step in the sequence), the necessary feedback signal is created to actuate cylinder *B*, which controls the high-pressure bleed. Cylinder *A* is actuated; this "ratchets" the sequencer to the next step in the cycle. This results in the output and sensor supply being switched to the next set of output holes on the distributor drum.

This type of sequencer provides a completely interlocked control circuit. This prevents power functions from occuring out of step due to false actuation of the various sensors in the system.

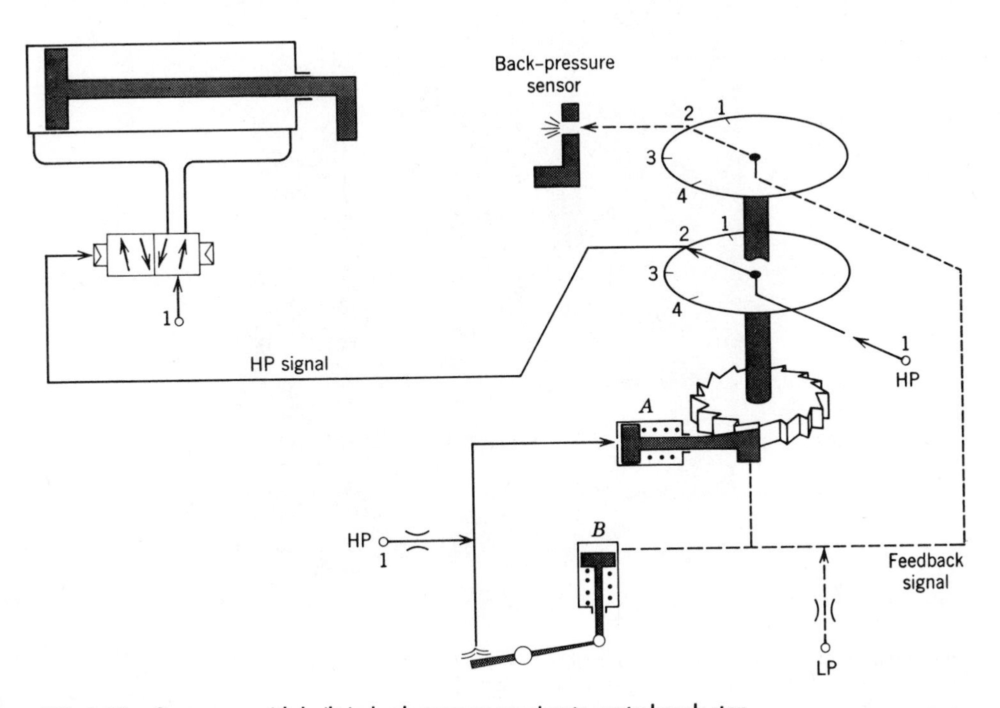

FIG. 4–23. Sequencer with built-in back-pressure sensing to control each step.

If a particular cycle requires less than the total number of steps in the sequence device, the unused output and sensing ports are capped. The capping causes the sequencer to skip the unused steps.

4-7 COUNTING DEVICES

Logic circuits can be designed to count; each count signal can be retained by the setting of a memory device. To count high numbers a large number of devices are necessary, and the display of the numbers is not easily accomplished. As a result pneumatically actuated mechanical counters are valuable in the construction of many automatic systems. Counters can be divided into the following types:

- Simple cumulative counters.
- Addition and subtraction counters that respond to input pulses.
- Predetermining counters which provide an output signal when a preselected number is reached. The output signal can be employed to stop a machine when a specific number of objects are counted.
- Predetermining counters with automatic reset can be employed to continuously count specific quantities; for example, the packing of products in specific quantities.

Pneumatic pilot light. (Courtesy Agastat, USA.) This pneumatic indicator displays green output when actuated and red when not. Because it protrudes from the panel, visibility from wide angles is possible. It operates from a high-pressure signal.

4-8 INFORMATION DISPLAYS (READ-OUT DEVICES)

When monitoring the output of automatic processes, read-out devices are required. A large variety of pneumatic and fluidic types of devices are available which are similar to electrical display devices in their operation. Fluid signals are converted into visible displays. A typical example is a display device which employs a brightly colored piston whose visibility is enhanced by magnifying its size with the addition of a lens. Audible readouts are easy to create with the addition of whistles or horns connected to the output circuit.

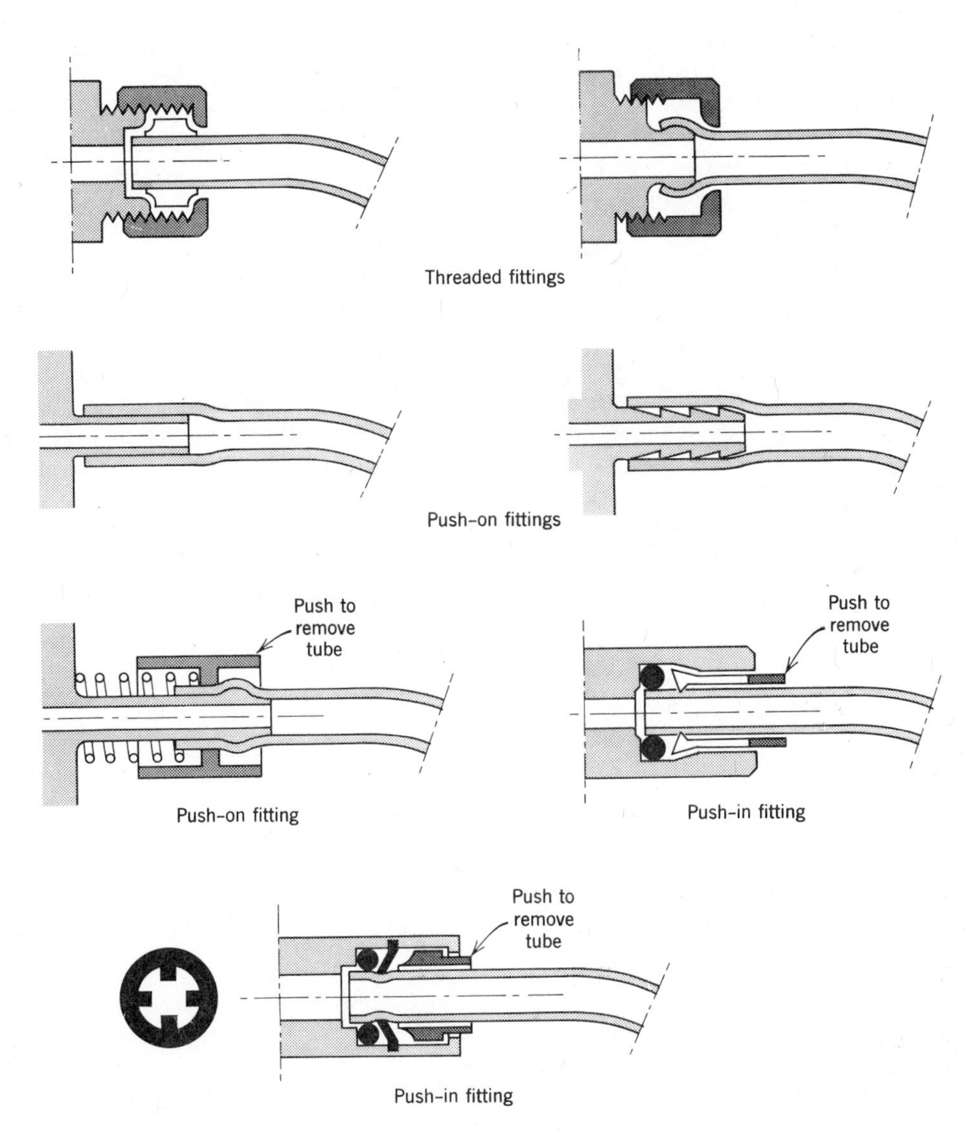

FIG. 4–24. Operating principles for various plastic tube fittings.

4-9 CONNECTING DEVICES

To interconnect logic and peripherial devices, tubing and fittings are required. Tubing is usually made from plastic materials to facilitate connections in a manner similar to electrical devices.

The various principles employed for fittings are illustrated in Fig. 4-24. Fittings are designed to provide a leak proof connection between tubing and devices.

The most common type of fitting is the compression type; however, with the continuing miniaturization of devices, it has become increasingly difficult to tighten the compression nuts when the fittings are closely spaced.

Push on fittings overcome the disadvantages above by employing barbs for high pressures and slip on styles for low pressures. Their disadvantage is that they create undesirable flow restrictions in the tubing.

A recent development in fittings combines the advantages of a one piece fitting, unrestrcted flow, and the quick connect and disconnect feature.

Similar to electrical connectors, multiple tube connectors are available which offer quick disconnection of the control enclosure or logic panel from the machine.

4-10 CONCLUSION

This chapter can be summed up by the following two points. The wide variety of peripheral devices available facilitates the implementation of the fluid logic technology. Sensors, for example, are frequently so important to the operation of a machine that they become the motivating factor for employing fluid logic control systems.

The well-established technology of electrical controls has aided the development of fluid-type peripheral devices. Evolutionary developments of fluid peripheral devices are expected to continue advancing this technology.

CHAPTER 5

THE SELECTION OF A FLUID LOGIC SYSTEM

In solving a digital controls problem, a choice must be made:

- In technology, the choice of fluid or electrical logic.
- In applying fluid logic, the choice of an appropriate system.

When seeking solutions to control problems, it is not practical nor even possible each time to obtain sufficient information or form valid comparisons and conclusions as to the most appropriate system or method to employ.

The purpose of this chapter is to define the basic criteria necessary in making the correct decisions that will solve the problems with a minimum of time and effort.

In the design of fluid logic devices, it is apparent that manufacturers have emulated electrical techniques that have been proven reliable in industrial automation applications.

To provide greater insight into the potential solutions and to help facilitate the choice of hardware and peripheral equipment, a definite comparison will be made between fluid and electrical logic techniques.

5-1 A COMPARISON BETWEEN FLUID LOGIC AND ELECTRICAL AND ELECTRONIC LOGIC TECHNIQUES

The classic analogy between fluid and electrical media is often employed:

- Electrical potential V corresponds to fluid pressure P.
- Electrical current I corresponds to fluid flow Q.
- Resistance to electrical current $R = V/I$ corresponds to flow restriction $R = P/Q$.
- The storage of an electrical charge Q' by capacity $C = Q'/V$ corresponds to the volume V' of a capacity of a pneumatic relay with a switching threshold pressure P, whose absorption capacity at switching is $C = V'/P$.

This analogy provides a limited understanding. Hence, a more practical comparison based on the physical differences encountered in control applications follows.

Figure 3-14 illustrated the differences between electrical and fluid logic switching tehnologies. In electrical logic as in fluid logic there are two basic approaches that must be considered:

- Nonmovable-part techniques include electronic solid-state and fluidic solid-state techniques.
- Movable-part techniques include electromechanical relays and standard pneumatic pressure logic techniques.

The table below illustrates the similarities between electrical and fluid control techniques from an application standpoint—output signal level, life expectancy, maintenance, and trouble-shooting.

	Nonmovable-parts systems	Movable-parts systems
Electric	Electronic logic solid state	Contact logic mechanical
Fluid	Fluidic logic solid state	Pneumatic logic mechanical
Output Signal level	Very low	High
Supply requirement	Precision regulation and filtration Filtered current and rectified voltage Filtered air supply and regulated pressure	Supply Media Electricity AC or DC Industrial compressed air supply
Life expectancy	Theroretically unlimited	Limited life Nominal 5 to 10 million contact cycles for electrical devices 10 to 100 million cycles for pneumatic devices
Maintenance and repair	Infrequent; specialist required	Periodically; only require basic knowledge

From their inception, nonmovable-part fluid logic systems have been compared with electronic logic systems.

The previous chart shows their simularities between these two solid-state technologies:

- Nonmovable parts.
- Very low output signal level.
- Well-regulated supply required.

In contrast to the simularities, very important differences separate these two technologies—Response time, efficiency, overall size, and such.

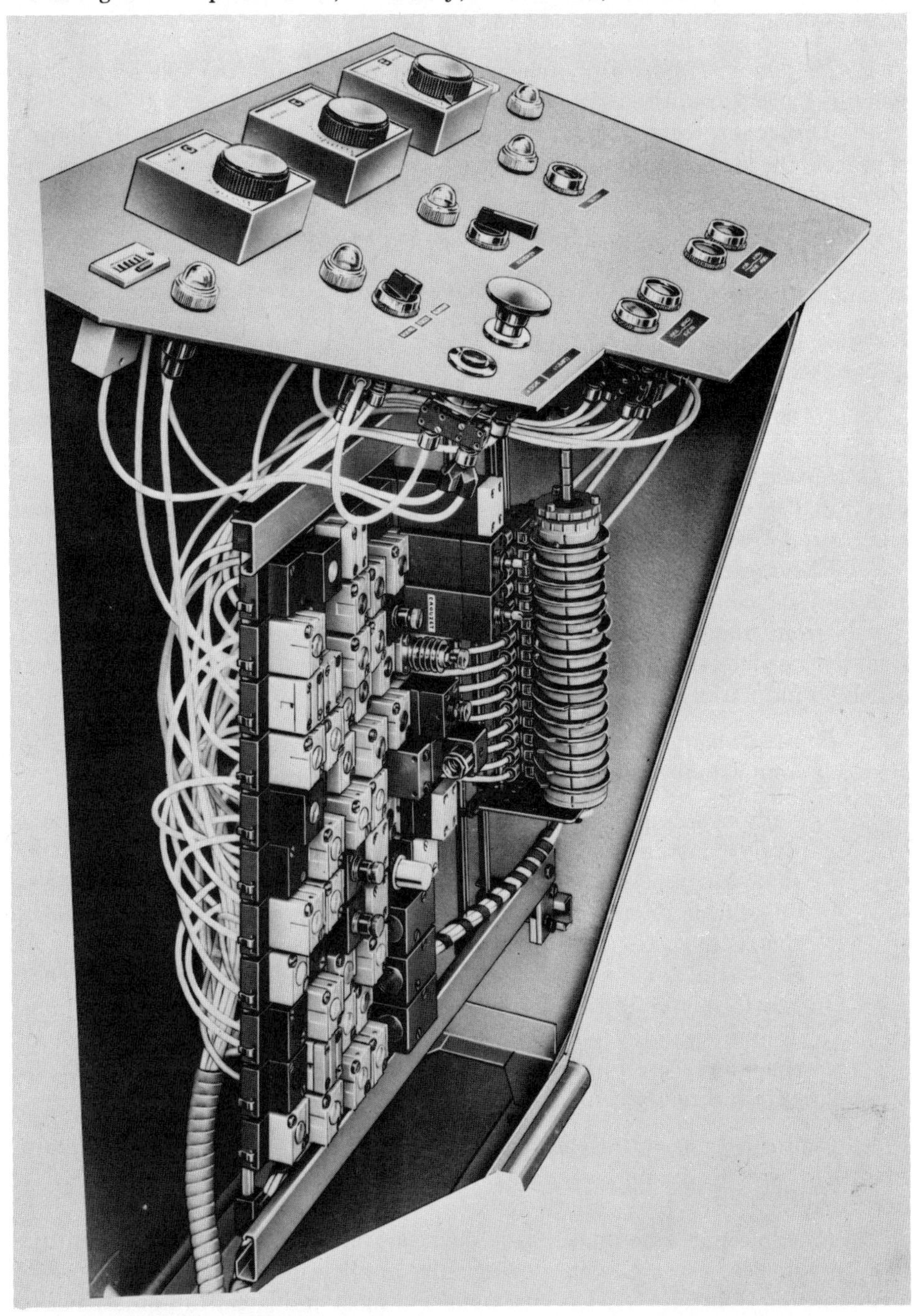

Pneumatic logic console (Courtesy Miller Fluid Power, U.S.A. and Crouzet, France.) This cut-away view shows how logic circuitry and peripheral items are assembled into an operator's console. The cover panel contains a count register, timers, switches, and push buttons. The cylinder below the top panel is a cam drum programmer for automatic sequencing of the logic system assembled on hangers and subplates to the left inside.

Figure 5-1 classifies the technologies in order of power consumption, the following characteristics: device response time, output power, and supply consumption.

Response Time. Only electronics is capable of the extremely high switching speeds required for computers, calculating machines, and such.

The other three technologies have response times 10^3 to 10^6 longer. However, these slower speeds are more than sufficient for most industrial applications.

Typical response times for these devices are as follows:

- Fluidic devices (all types) have a nominal response time of 1 to 3 ms.
- Movable-part, electrical, and pneumatic devices are slower in switching speed because of the inertia of the movable parts. The nominal response of 10 ms is typical for most movable-part devices.

Power Output. The power output of electronic and fluidic devices is very low (see Fig. 5-1 for data). In most applications an amplifier or power relay is required for operating industrial power devices.

Electromechanical devices (relays) provide output power signals equivalent to 100 times greater than the above devices. Moreover, output power signals from pneumatic devices can be up to 1000 times greater than solid-state devices, electronic or fluidic. These high-power output levels are directly usable in industrial applications.

Power Consumption. This often overlooked characteristic of logic devices is an important factor in differentiating between technologies.

- Nonmovable-part technologies (electronic and fluidic) have a relatively constant power requirement, in that they consume about the same amount whether the device is in the "ON" or "OFF" state.
- Pneumatic devices consume power at a rate directly proportional to the cycling rate (consumption only during switching).
- Electrical devices (relays) consume only when the solenoid is energized; this condition may be intermittent or continuous, and most systems employ circuits which require both conditions. Therefore, the power consumption figures illustrated are averages which are only indirectly connected to the cycling speed of the devices.

Figure 5-1 classifies the technologies in order of power consumption, from the lowest to the highest, electronic, electromechanical, and nonmovable-part fluid logic.

Movable-part pneumatic logic devices are by far the economical of the four technlogies, especially in slow cycling applications—security systems or door opening controls. This is also true for most industrial automation systems.

Pneumatic movable-part logic devices with their 1000:1 power output ratio compared to electronics and fluidics are by far the most efficient in the operating range of one cycle per second or less. Above this cycling speed they may equal or exceed the power consumption of electronic and fluidic devices.

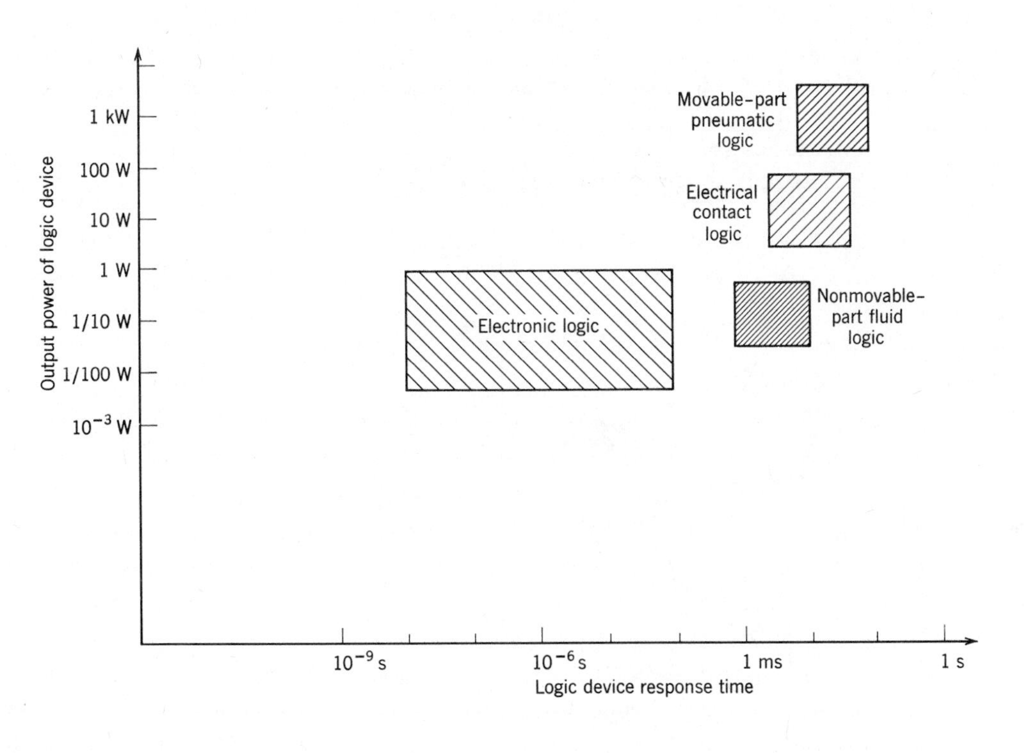

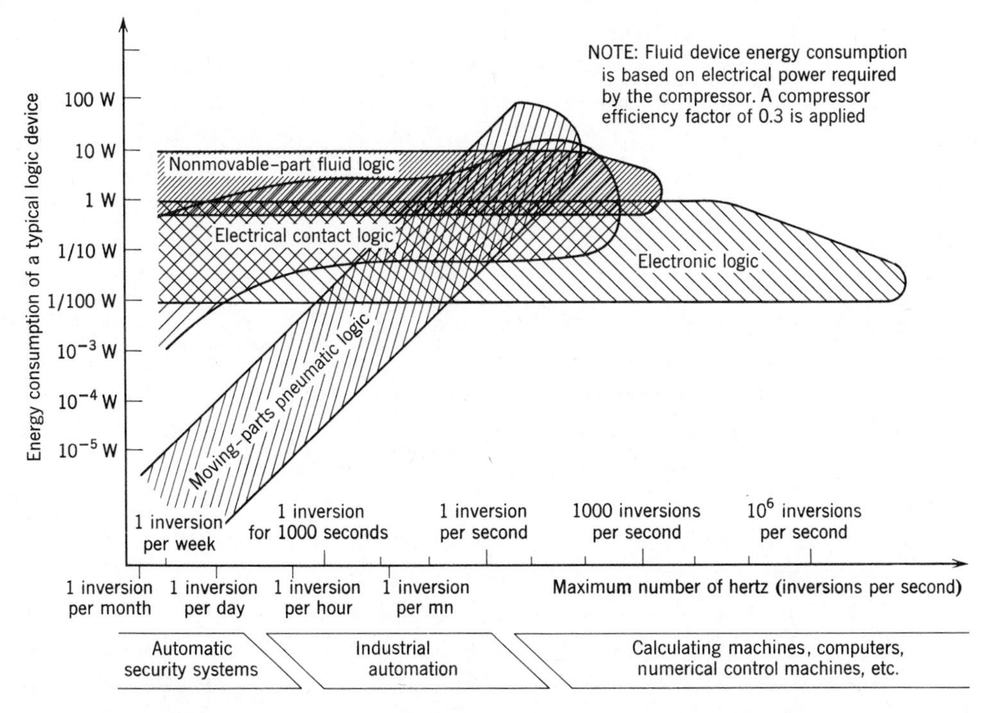

FIG. 5–1. Technological comparison between electric, electronic, and fluid techniques.

Another important characteristic of logic devices is the method required to create the memory function. Figure 5-2 describes the various methods and classifies them according to type:

- Memories that are maintained even when there is an interruption in the supply to the device—electrical or fluid supply.
- Memories that lose state—reverting to "O" when an interruption in the supply to the device occurs.

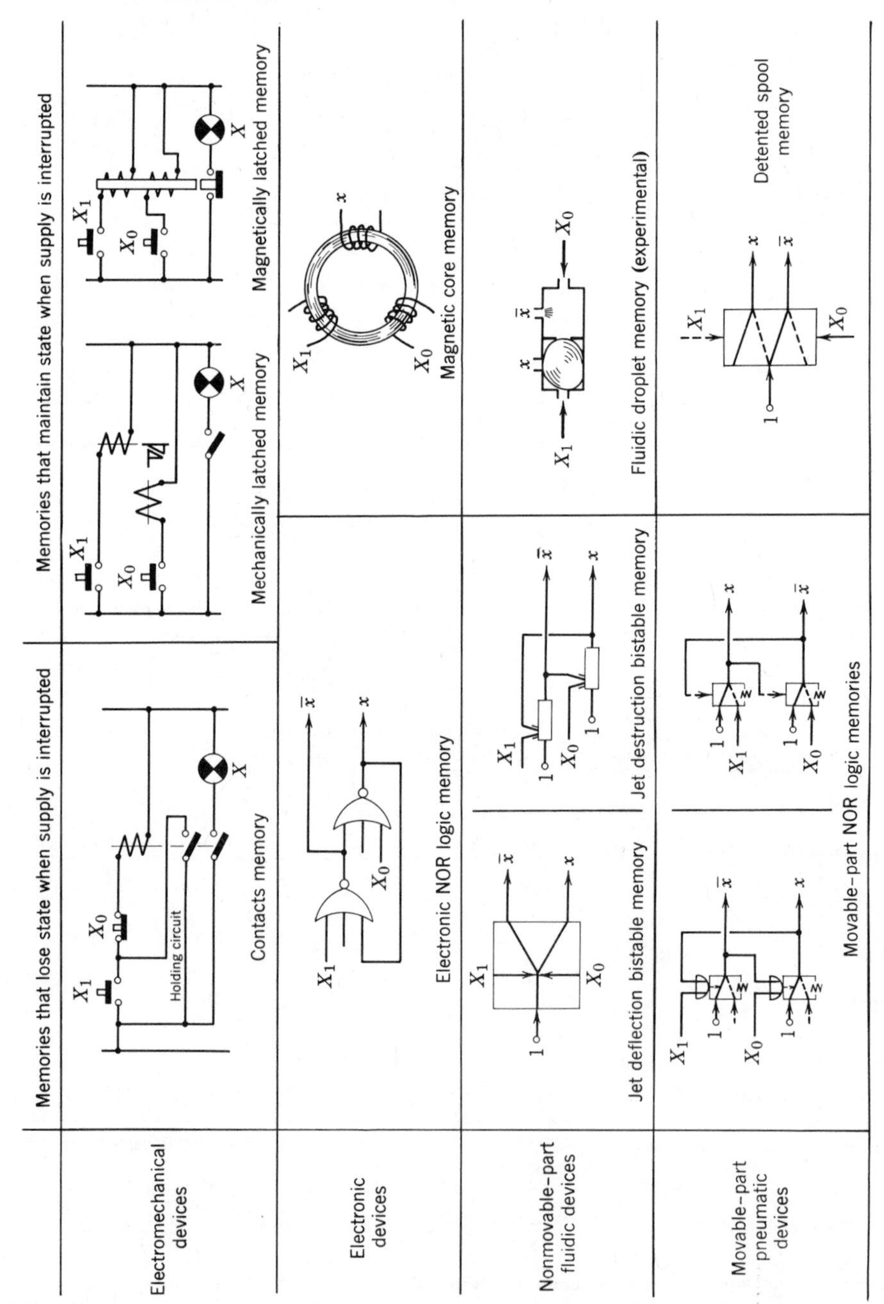

FIG. 5-2. Memory functions created with the various technologies.

Electricians are familiar with the problems encountered in control systems that employ memories that lose state when a power supply interruption occurs. The machine stops at a particular step in the cycle sequence and all the memories switch back to the "O" state, which may be the cycle start mode.

Memories that employ holding circuits cause this problem, since holding circuits release when the supply is interrupted.

The development of mechanically latched memories and later magnetically latched memories overcame this problem of losing state, because the retention of a state (0 or 1) is not dependent on the electrical supply.

Two types of memory devices are available for electronic systems:

- Memory functions created with holding circuits, employing, for example, NOR logic devices.
- Magnetic core memories retain state even after power is removed. Their application is universal in electronic computers.

In movable-part pneumatic devices described in Chapter 3, retained memories are built with spool valves. (see Fig. 3-18).

In complex applications with restarting problems, memories that retain their state independent of supply state are recommended over memories that are dependent on holding circuits.

Nonmovable-part fluidic devices do not currently offer memory devices which are independent of supply state. Only the use of retarding circuits can overcome this problem for jet destruction devices and jet deflection devices.

As illustrated in Fig. 5-2, "droplet-type memory" devices are currently under development. They operate on the principle of pushing a liquid droplet from one chamber to another by the force of piloting pressures. Either output x or $\bar{x}$ is closed or open, according to the position of the droplet. Since the position of the droplet will be maintained even when an interruption occurs, this method may be the answer to the memory problem for fluidic type controls.

It is of importance to mention that all of the memory devices noted as being unable to retain state as a result of an interruption in supply will always return to a specific state when supply is removed and will maintain that specific state when supply is re-established.

5-2 THE CRITERIA THAT DETERMINES SELECTION BETWEEN FLUID, ELECTRICAL, AND ELECTRONIC TECHNOLOGIES

In Chapter 1 we described basic reasons for employing fluid logic controls: the control of fluid power output devices; controls employed in an explosive environment; and such.

Now that we have described the basic fluid logic devices and the peripheral equipment, we can define fully the criteria for selecting fluid logic controls in contrast to employing the more traditional electrical controls.

Figure 5-3 illustrates a hypothetical system of compatible devices.

All automation systems require inputs from sensors and other peripheral equipment. The logic circuit interprets these input signals and then sends appropriate control signals to the output devices and indicators. A system consisting of the greatest number of compatible components from the inputs through to the outputs will minimize the number of interface devices required, which will provide a system of minimum complexity, lower the cost of components and construction, and more importantly offer a more reliable system.

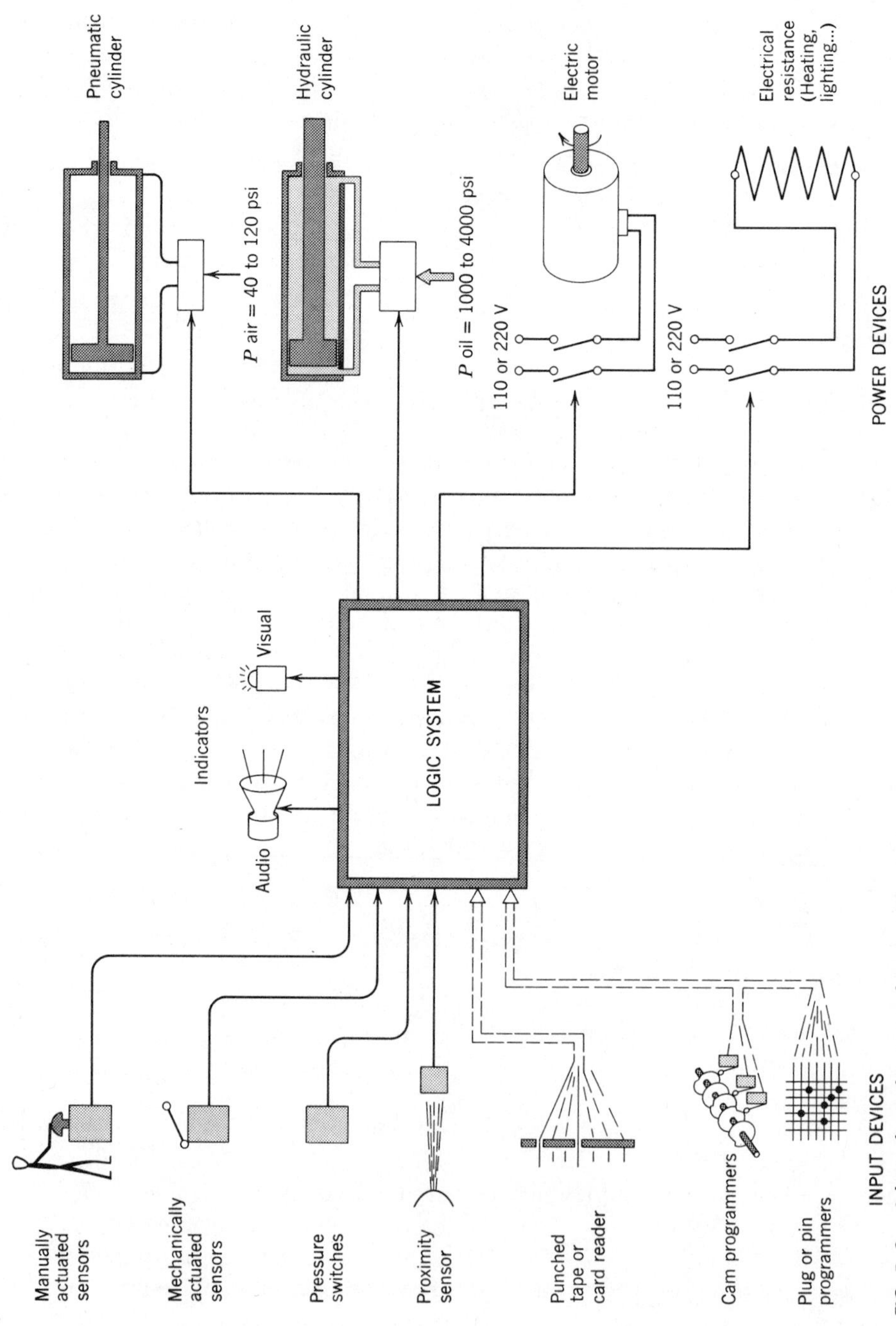

FIG. 5–3. A hypothetical system of compatible devices.

If the control requirement is primarily for electrical devices, electric motors, resistance devices (lighting or heating), and such, then the use of electrical or electronic controls is preferred.

In contrast, if the power devices are primarily fluid power devices, pneumatic or hydraulic, fluid logic controls will provide the best overall method of control.

Between these two extremes are the countless applications that require fluid power and electrical power devices. In these cases careful consideration should be given to the number of interface devices required for each method of system control. The final selection should be based on the method which minimizes the number of solenoid valves and air-electric switches required.

The technology of fluid logic controls eliminates the need for the traditional electropneumatic and electrohydraulic control systems and thereby provides the means to design and construct systems employing only fluid power devices.

This approach would allow the use of pneumatic pressure switches instead of pressure electrical switches with their inherently limited life expectancy.

Hydraulic devices can be controlled by air piloted hydraulic valves which are manufactured by all major hydraulic valve companies and have been available for a number of years.

The compatible system concept just described is important. However, a selection of system controls can be influenced by other requirements; the following are typical cases.

- Control systems operating in *explosive environments* employ fluid logic controls whenever possible. Electrical components create a danger of sufficient magnitude to require their installation in specially sealed explosion proof enclosures. These enclosures are expensive and require a larger area for installation. The problem of operating machines in explosive environments is of particular importance in all chemical, petroleum, and mining industries, and also industries employing or bottling gases or explosive liquids.
- *Complex automatic control systems* consisting of thousands of individual logic devices require ultrafast switch speeds (response time) which only can be satisfied by electronic devices. Typical examples of control systems requiring ultrafast response times are: computers, and their peripheral equipment, and numerical control systems. Fluidically controlled calculating systems have been manufactured. Some numeric control machines employing fluidic devices have also been manufactured. However, the trend towards this type of control does not seem to be significant at this time. As a result of its inherently fast response times, electronics is the only technology capable of satisfying the needs of calculating systems at present.
- Speed of response is also of great importance when interconnected machines are separated by long distances. In this type of situation, the *propogation speed of the signal* is of major importance. When comparing fluid logic with the electrical logic, Fig. 5-1 illustrates the fact that the response time is equivalent for movable-part and non-

movable-part devices. However, Fig. 5-1 does not mention that the signal propogation speed to and from the various electrical and pneumatic devices is quite different.

- An electrical signal is propagated at the speed of light, 286,000 miles/sec.
- A fluid signal cannot be propagated faster than the speed of sound, approximately 1100 ft/sec.

Pneumatic programmed lathe. (Courtesy Techne, Great Britain.) These views illustrate the adaptation of pneumatic programming to lathe machining. The horizontal array of pushbuttons on the above panel are used by the operator to make setup adjustments before switching over to automatic operation programmed by the rotating drum just above them. The cabinetry above the machine contains low-pressure logic as well as high-pressure power circuitry.

The speed ratio is therefore enormous: $\simeq 10^6$ in favor of electricity. In automation systems that require the sensors and the power devices to be located at a distance from the logic controls, and where high-speed operation is of importance, the electronic or electrical control systems should be employed to eliminate the potential problem of signal propagation speed. For example, the central control of an entire factory's machines or the control of a long production line require that the control system employs electrical controls. In contrast, the operation of filling valves on large tankers may be controlled by pneumatics, because a control line of a thousand feet in length will propagate the signal to the control valve in a second or so. This response speed is compatible with the needs of this application. If a power system uses fluid power devices and is electrically controlled, the response of the solenoid valves must be taken into consideration. The time required for the solenoid to respond to the electrical signal is a nominal 100 ms. With the response speed of solenoid valves in mind, it becomes apparent that fluid logic control is more advantageous overall than electrical controls, if the entire system is not spread over long distances or large areas. For example, an automatic packing machine employing pneumatic cylinders can operate at higher speeds with a fluid logic control than with an electrical logic control.

- *The selection of input devices and other peripheral equipment*

  Among the peripheral equipment that may lead to the utilization of electrical and electronic devices are the following:

  - Plug programmers of which the fluid equivalent are larger and more costly.
  - Magnetic tape recorders are unknown in the fluid device technology although punched tape readers are in use.
  - Visual information displays are available in fluid versions; however, they are not as versatile as their electrical equivalents and their cost is usually higher.

  Among peripheral equipment that is more desirable in its fluid versions we mention:

  - Sensors that are required to operate in wet or dusty environments are usually more reliable and tolerant of these environments if they are fluid devices.
  - Some sensing functions and applications are not feasible with electrical devices. Among these are the sensing of nonmetallic devices that are in an environment that would either give false signals to electric eyes because of stray light or reflection, or in applications that would blind "electric eyes" due to dust, smoke, or steam.
  - The versatility of back-pressure sensors is not equalled by any electrical sensor, especially when the sensor consists of nothing more than small diameter hole in a die or fixture to detect the presence of an object.

- The advantages of the pneumatic pressure switch have been discussed earlier. Fluid type liquid level sensors are more reliable than their electrical equivalent in end applications.
- Sensors designed to detect extremely light objects for gauging purposes are more reliably sensitive and easier to apply than their electrical equivalent.

5-3 THE CRITERIA OF SELECTION AMONG THE VARIOUS FLUID LOGIC CONTROL SYSTEMS

When the selection of controls has been narrowed down to the technology of fluid logic, it is then important to select the most suitable system for a particular problem.

Before making the selection above, one should take into consideration the compatibility of the required operating pressure of the system, and the input and output devices. The appropriate operating pressure should minimize the number of interface devices. The types of input and output are a guide to the most desirable system operating pressure.

Two examples illustrate this point. If a power function is to be accomplished by three pneumatic cylinders equipped with pressure switches, the utilization of high-pressure movable-part logic is the most compatible with the overall system.

In contrast if a control system must operate two valves according to a program controlled by a five-channel punched tape reader, a low-pressure, possibly nonmovable-part system would be the best choice. This selection would necessitate only two output amplifiers as opposed to the high-pressure system which would require five input amplifiers.

To facilitate the selection of the most appropriate fluid logic control system we have transposed the technological classifications discussed in Chapter 3 to create Fig. 5-4.

On the operating pressure scale, three practical pressure ranges must be considered:

Standard Industrial Pressures Range (40–150 psig). The utilization of a supply pressure within this range for logic devices requires the amplification of low-pressure sensor outputs. However, the direct operation of power devices is possible directly from the logic output.

Pneumatic Instrumentation Pressure Range (3–30 psig). A fluid logic system operating within this range is directly compatible with pneumatic instrumentation systems, and power supplies.

Low-Pressure Range (0.2–10 psig). Fluid logic systems operating within this range can directly utilize the output of sensors operating on the same supply pressure as the logic. Output interfaces are necessary to amplify the logic output signals to actuate power devices.

On the flow passage diameter scale, the degree of filtration required for the three types of fluid logic devices has been discussed in section 4-2. The

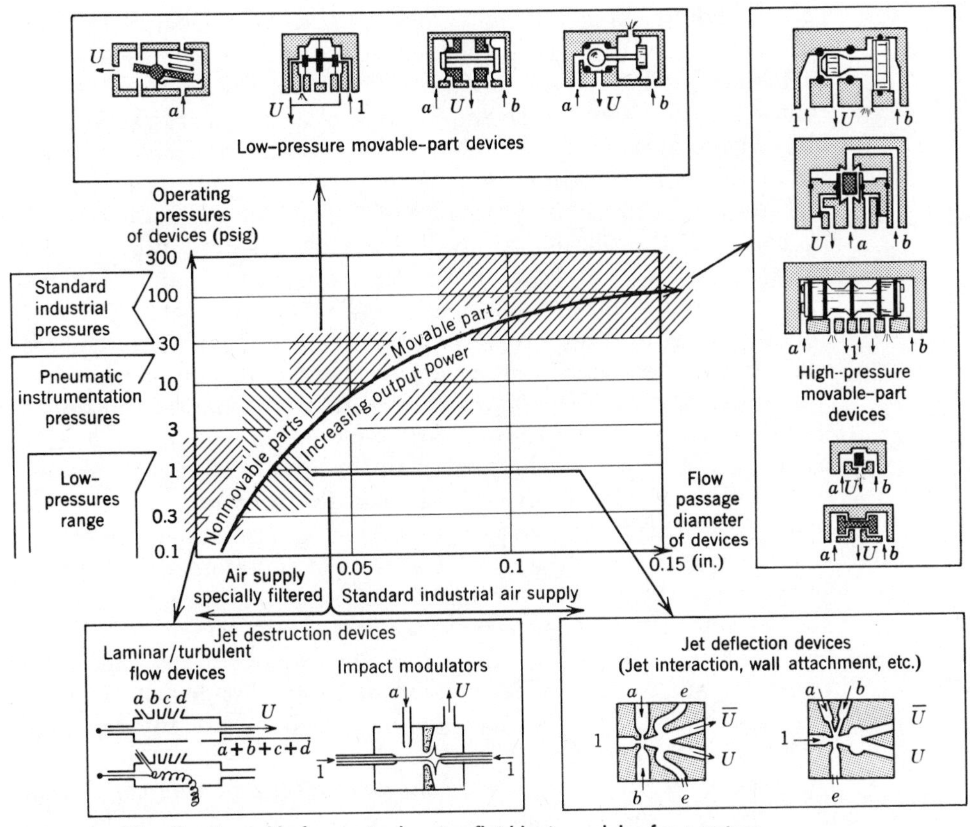

FIG. 5–4. Classification table for use in choosing fluid logic modules for a system.

smaller the diameter of the flow passages, the greater the need for high filtration and water and oil removal from the supply.

Other considerations may influence the selection among the various fluid logic systems. Here are the important ones:

- *The air consumption of logic devices* is a point discussed in section 5-1. Figure 5-1 illustrates the fact that movable-part logic devices consume the least amount of air as long as the cycling rate is smaller than 5 Hz. Nonmovable-part fluid logic devices become more economical on air consumption, for high cycling rates only.
- *The life expectancy of logic devices* is a very important criteria for selection between electronic (theoretically unlimited life expectancy), and electromechanical devices (5 to 10 million cycles life expectancy). However, within the technology of fluid logic this criteria is not as important in the selection of a system. The life expectancy of fluidic logic devices is theoretically unlimited as with electronic devices. The life expectancy of movable-part logic devices can usually exceed 50 to 100 million cycles. This factor becomes important only when high-speed operation and long-production runs are required that may exceed the normal life expectancy of movable-part logic devices.

- *The availability of maintenance personnel* to trouble-shoot the various technologies can become a decisive factor in system selection. A particular problem exists for small factories; the number of machines may not warrant the employment of a specialist to service the fluid logic equipment. Electrical and electronic control equipment requiring a specialist for servicing has impeded the acceptance of this equipment in industry because of the lack of trained personnel. In contrast, the simplicity, ruggedness, and safety provided by movable-part logic systems minimizes the need for trained maintenance personnel for trouble-shooting. The majority of the time a maintainer with only a basic understanding of pneumatic devices can service adequately the systems. However, in nonmovable-part fluid logic the problems are not readily visible to the untrained maintenance man; therefore, maintenance and trouble-shooting usually requires a specialist with a knowledge of not only the operation of the individual devices, but also the precautions needed to ensure that the system does not become contaminated while it is being serviced, or to eliminate the sources of contamination if that is what created the need for service.
- *The response times of devices* were discussed in section 5-1. When high-speed control systems are required, nonmovable-part logic devices are usually more efficient than movable-part devices, as is illustrated in Fig. 5-1. Nonmovable-part systems are the practical choice when the cycling rate exceeds 5 to 10 Hz.
- *The cost of devices, and their size, mounting, and interconnecting methods,* and such are all very important points to consider also before making a system selection. We leave these details to the reader primarily because of the rapid evolution in this field. New systems appear almost on a regular basis usually offering many new practical advantages. Whenever possible the reader should compare *systems,* not individual logic devices. He will then be able to take into account the peripheral equipment which is available for each system.

5-4 CONCLUSION

The selection between electrical and fluid logic control technologies is relatively easy after some basic knowledge of each is acquired. However, to make the choice between the most appropriate fluid logic system requires more investigation and experience.

Aside from the criteria discussed, we hope that the application problems in Chapter 8 will assist the reader in making the best decision possible when seeking the solution to industrial automation control problems.

CHAPTER 6

BOOLEAN ALGEBRA REVIEW

Chapter 2 briefly explained the basics of Boolean algebra necessary for understanding the operating principles of fluid logic devices. This chapter reviews and further expands on the concepts already introduced, so that the reader will be better able to implement the Boolean algebra manipulations important in logic circuit design.

6-1 DEFINITIONS

6-1-1 Binary Devices

The operating principle of a binary device is characterized by two states that are complementary and opposing. The states are typified by electrical contacts, pneumatic valves, transistors, and such.

6-1-2 Binary Variables

Two values are assigned to the states of a binary device, 1 and 0. The assignment of the values 1 and 0 to the states of a device are completely arbitrary. For example, the states of pneumatic devices are characterized by the 1 state when a particular port of the device is under pressure and by 0 state when a port is connected to exhaust (atmosphere). The states of an electrical device are characterized as 1 when a wire is charged and 0 when it is discharged.

Electrical devices are characterized by the 1 state when active and the 0 state when inactive. The 1 and 0 do not represent any numerical values; their only significance is that they create a distinction between the two binary states of a logic device.

Figure 6-1 illustrates the operation of typical electrical and pneumatic devices and their binary states. Upper case letters E and P denote binary devices, lower case letters such as e, $\bar{e}$, p, $\bar{p}$ indicate the output states of these devices. All of these are binary devices.

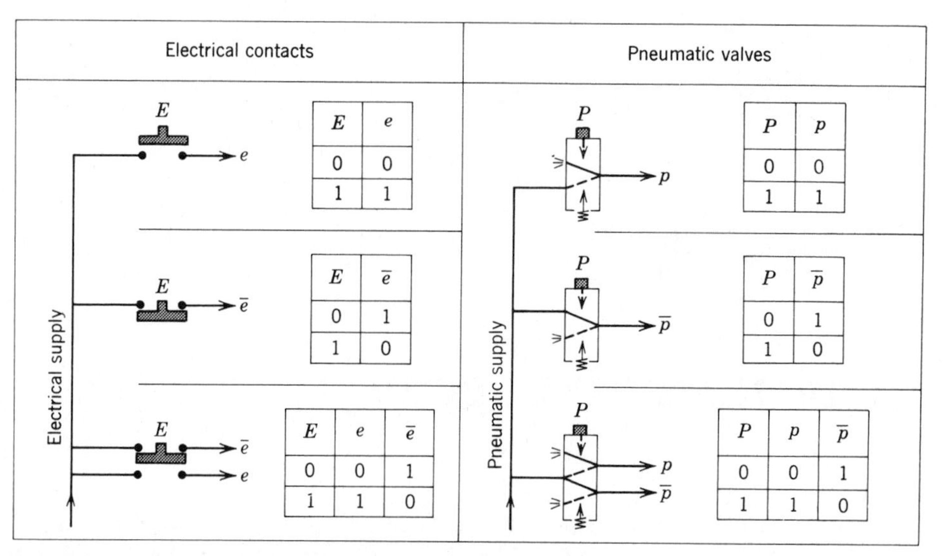

FIG. 6–1. Digital devices (electrical and pneumatic) and their binary variables.

6-2 BOOLEAN ALGEBRA—BASIC MANIPULATIONS

In designing an automated system, Boolean algebra provides the means of mathematically expressing the control requirements of a machine. The general method is to connect the inputs by their operating symbols, which represent their operating requirements.

The operating symbols employed are sometimes the same as those employed in basic algebra; however, there is a fundamental difference in their meaning when applied to Boolean algebra. In Chapter 2, we introduced the YES, NOT, OR, AND functions. We now expand on these further.

6-2-1 YES Function (Logic Equality)

This function represents a situation where two binary devices have the same constant values. If from

$a = 1$ results $b = 1$

and from **thus $a = b$**

$a = 0$ results $b = 0$

An example of this would be a sensor which provides an output signal when an object enters its range. There is always an equality between the object being sensed and the output signal.

6-2-2 NOT Function (Logic Inversion)

If two binary devices have constantly opposing logic values they are connected by a NOT function. If from

$$a = 1 \quad \text{results} \quad b = 0$$

and from

$$a = 0 \quad \text{results} \quad b = 1$$

thus $\boxed{a = \bar{b}}$ or $\boxed{b = \bar{a}}$

In the NOT function a and b are known as inverse or complementary variables.

A typical application is illustrated in Fig. 6-2.

- When sensor A is actuated (state 1), the fill valve V is closed (state 0).
- Inversely when A is not actuated (state 0), V is open (state 1).

Valve V is connected to the sensor by a NOT function $\longrightarrow V = \bar{a}$.

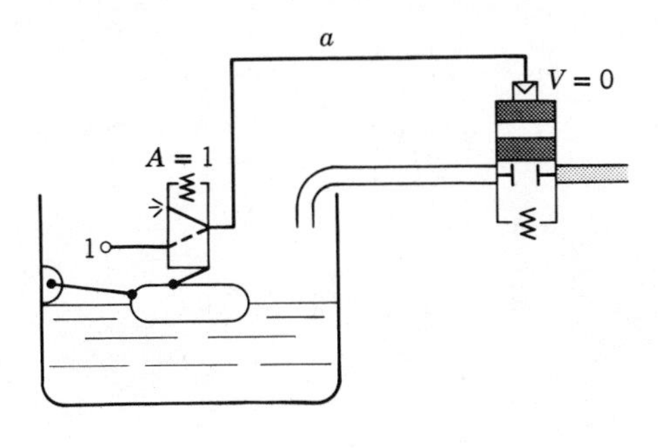

A	a	$\bar{a}$	V
1	1	0	0
0	0	1	1

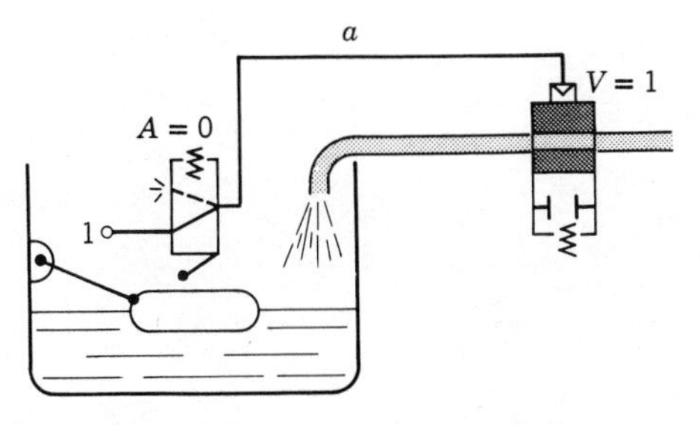

FIG. 6–2. Illustrates an application of the NOT logic function. When sensor valve *A* is NOT actuated, it provides an input pilot signal to open fill valve *V*.

Note. The bar symbol over a variable inverts it to its complementary value. If

$$a = 1, \qquad \bar{a} = 0$$

and if

$$a = 0, \qquad \bar{a} = 1$$

can also be stated as:

$$\boxed{a = \bar{\bar{a}}}$$

6-2-3 The AND Function (Logic Product)

The AND function expresses an output U which $= 1$ when both variables a AND $b = 1$, and only when this requirement is satisfied.

$$\boxed{U = a \cdot b} \qquad U = a \text{ AND } b$$

Binary variables a and b each have two states which are connected and represent $2^2 = 4$ possible states. The b can $= 1$ and 0 values when $a = 1$ or when $a = 0$; this provides four possible combinations.

The truth table in Fig. 6-3 illustrates the four possible combinations and the relationships to the states of $U = a \cdot b$.

Note. The arithmetic product of a and b equals the same value as the logic product $U = a \cdot b$.

a	b	$U = a \cdot b$	
0	0	0	$0 \times 0 = 0$
0	1	0	$0 \times 1 = 0$
1	0	0	$1 \times 0 = 0$
1	1	1	$1 \times 1 = 1$

Truth table for function $U = ab$

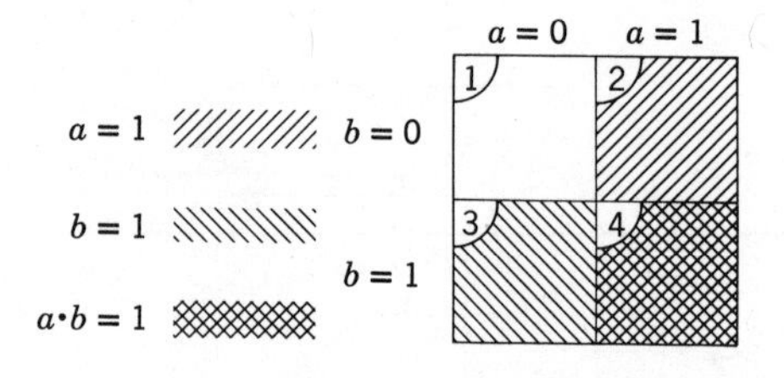

Karnaugh maps for two variables

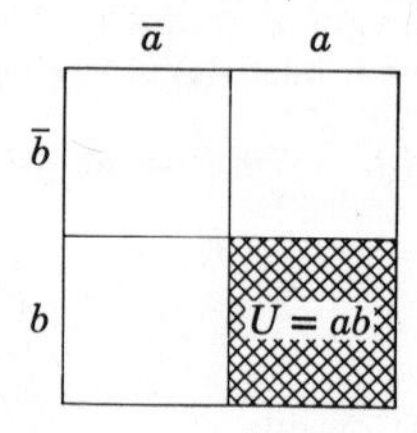

FIG. 6-3. Illustrates the AND function $U = ab$, logic product.

Figure 6-3 also illustrates the truth table information in the form of two Karnaugh maps, both of which represent the same information on the possible states of $U = a \cdot b$. Each cell of the maps equals one of the four states. The cross-hatched cell in each map represents the only combination of a and b which equals 1. The AND function is also known as the intersection or coincidence function on Karnaugh maps.

The only difference between the two maps is that the second map is a simplified version of the first. The representation of the possible combinations of variables is very easy and practical when the Karnaugh mapping method is employed.

Other methods of representation are available, such as the Euler method. Some of the other methods are even more detailed than the Karnaugh maps; however, they are more difficult to interpret.

6-2-4 AND Functions with Three or More Variables

The two input AND function $U = a \cdot b$ is the basic combination. Once it is understood, functions with 3, 4, or n variables may be employed; $U = 1$ only when all variables, 3, 4, or n, all equal 1 simultaneously.

$$U = a \cdot b \cdot c \cdot d \cdot e \; . \; . \; . \text{ etc.} \qquad \text{or} \qquad U = abcde \; . \; . \; . \; ,etc.$$

The Karnaugh mapping method can be extended to represent any number of input variables for the AND function. Figure 6-4 illustrates $U = abc$ which is a three-input AND function. Since each input has two states, there are $2^3 = 8$ possible combinations of these three variables.

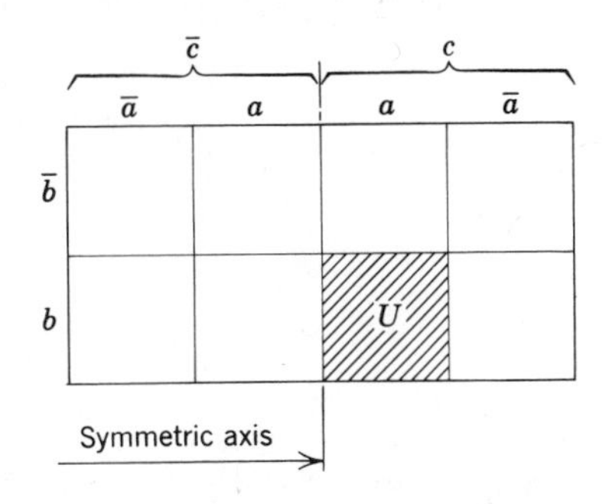

FIG. 6–4. Illustration of function $U = a\ b\ c$.

A map consisting of eight cells will be necessary. First draw one map of four cells representing the four states of variables a and b, and then draw a duplicate or mirror image of the map, directly adjoining it, to the right. This is known as binary expansion. Each of the maps containing four cells will correspond to one state of c, c and $\bar{c}$.

The four possible states of a and b can be represented in state c or $\bar{c}$. The diagonally lined cell corresponds to variables c, b, and a; therefore,

$$\boxed{U = abc}$$

Note. When adding a fourth variable to the map, the number of cells will need to be symetrically doubled again. The additional set of eight cells can be added to the right or under the original set. Each value of the fourth variable is given to a set of cells. This method of extension can be repeated to accommodate n variables. Each set of 2^n cells correspond to the two combinations of n variables.

6-2-5 OR Function (Logic Sum)

Output $U = 1$ when one (or both) of the variables a OR b is equal to 1. In Boolean algebra this is represented by

$$\boxed{U = a + b} \qquad (U = a \text{ OR} b)$$

Figure 6-5 illustrates the truth table for all four possible states of the combinations a and b. Note that the arithmetic sum equals the logic sum $U = a + b$, except when a and b both equal 1. In binary logic $1 + 1 = 1$ since the number 2 is not considered.

The Karnaugh mapping method can be employed to graphically represent the OR function, just as with the AND function. Each lined cell is where $U = 1$. Any empty cell is where $U = 0$. With a two input OR: $U = 1$ when $a = 1$ OR $b = 1$. Therefore,

$$\boxed{U = a + b}$$

Truth table

a	b	$U = a + b$	
0	0	0	0 + 0 = 0
0	1	1	0 + 1 = 1
1	0	1	1 + 0 = 1
1	1	1	1 + 1 = 1

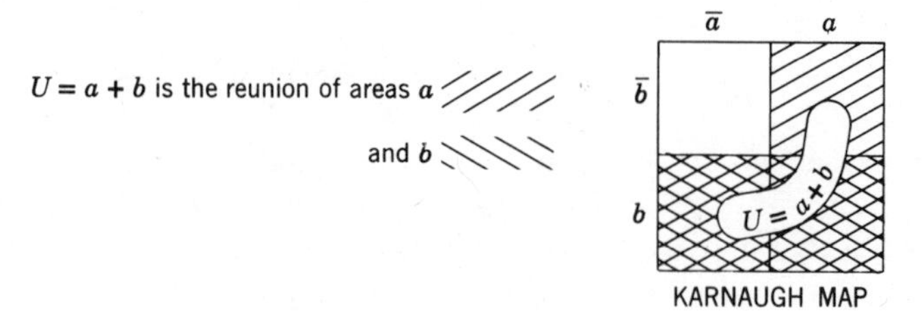

FIG. 6–5. Truth table and Karnaugh map illustrating the OR function, $U = a + b$.

The OR function is also known as the reunion function, because $U = a + b$ is a reunion of a on one side and b on the other. We have just described a two input OR function, but we can just as easily extend the number of variables to n inputs.

The OR function only requires that one of the inputs equal 1 for the output to equal 1. Figure 6-6 illustrates a Karnaugh map representing a four input OR function.

$$U = a + b + c + d$$

All lined cells represent $U = 1$. The empty cell represents $U = 0$.

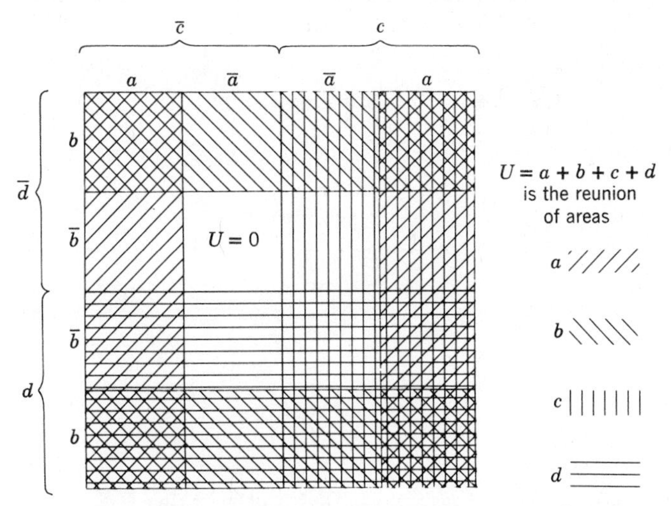

FIG. 6–6. A Karnaugh map illustating $U = a + b + c + d$, a four-input OR function.

6-2-6 The Exclusive OR Function

The OR function described previously is known as an inclusive OR function, because it includes the logic product $ab(a = 1 \text{ AND } b = 1)$ represented by the square with cross-hatched lines. This is the most common type of OR function. There is another type of OR function known as the exclusive OR function which excludes logic product $a\ b$.

The exclusive OR function represents $U = 1$ when $a = 1$ AND $b = 0$ OR when $b = 1$ AND $a = 0$. $U = 1$ when ever one input equals 1 and the other input equals 0. Figure 6-7 illustrates the Karnaugh map for the exclusive OR. The U is represented by the diagonally lined squares $U1$ and $U2$. In Boolean algebra terms $U = U1 + U2$. The $U1$ is the intersection of areas a AND $\bar{b}$ $\rightarrow$ $U1 = a\bar{b}$; $U2$ is the intersection of areas b AND $\bar{a}$ $\rightarrow$ $U2 = b\bar{a}$.

$$\boxed{U = a\bar{b} + b\bar{a}} \quad \text{also written as } U = a \oplus b$$

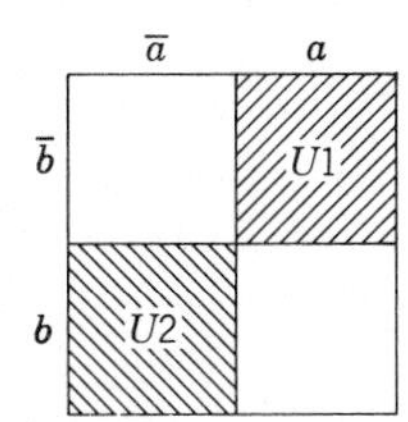

FIG. 6–7. Karnaugh map illustrating the exclusive OR function: $U = \bar{a}b + a\bar{b} = a \oplus b$.

Note. In general the inclusive OR function is applied unless the exclusive OR function is mentioned.

6-3 THE CHARACTERISTICS OF BOOLEAN ALGEBRA

We have now covered the basic operations of Boolean algebra: equality (YES), inversion (NOT), product (AND), and sum (OR). As in basic algebra, these operations can be combined into equations which we can manipulate, by following the various rules which shall be discussed.

6-3-1 Some Characteristic Logic Relationships

The relationships provide a means of simplifying logic equations.

Negations	Products	Sums
$\bar{\bar{a}} = a$	$1 \cdot 1 = 1$	$1 + 1 = 1$
$\bar{1} = 0$	$a \cdot 0 = 0$	$a + 0 = a$
$\bar{0} = 1$	$a \cdot 1 = a$	$a + 1 = 1$
	$a \cdot a = a$	$a + a = a$
	$a \cdot \bar{a} = 0$	$a + \bar{a} = 1$

6-3-2 Commutative, Associative, Distributive Rules

They are similar to the equivalent rules in basic algebra. A brief review:

commutative	associative
$ab = b \cdot a$	$a \cdot (b \cdot c) = (a \cdot b) \cdot c$
$a + b = b + a$	$a + (b + c) = (a + b) + c$

distributive
$ab + ac = a(b + c)$
$(a + b)\ (c + d) = ac + ad + bc + bd$

6-3-3 De Morgams' Theorems

The NOT function can be applied to a complete expression.

$$U = \overline{a \cdot b} \quad \text{or} \quad X = \overline{a + b + c} \quad \text{or} \quad Y = \overline{(a + b)\bar{c}} \text{ etc.}$$

De Morgans' theorem provides for the transformation of inverted expressions. Figure 6-8 illustrates two of De Morgans theorems by means of Karnaugh maps.

THEOREM 1

The complement of an OR function is the AND function created with the complements of its variables

$$\text{if } U = a + b \rightarrow \overline{U} = \overline{a + b} = \bar{a} \cdot \bar{b}$$

THEOREM 2

The complement of an AND function is the OR function created with complements of its variable

$$\text{if } U = a \cdot b \rightarrow \overline{U} = \overline{a \cdot b} = \bar{a} + \bar{b}$$

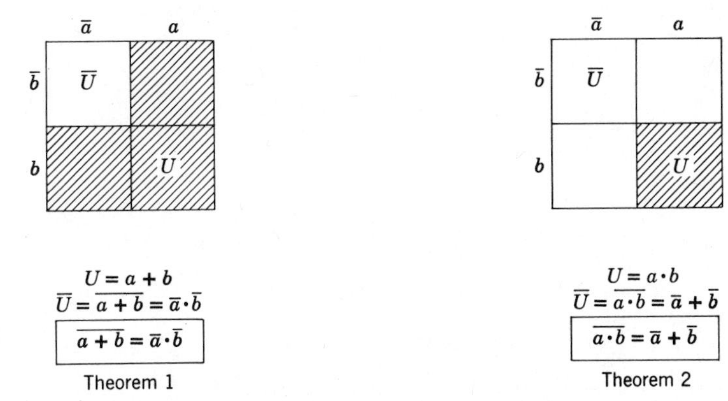

FIG. 6–8. Graphic illustration of De Morgan's theorem.

RULES OF COMPLEMENTS

The complement of a logic expression is obtained by complementing each of the terms of an expression and by changing the connecting signs (the + symbol to the • symbol and the inverse).

EXAMPLES

$$\overline{(a + b)\bar{c}} = \overline{a + b} + c = \bar{a}\bar{b} + c$$

$$\overline{\bar{a}b + \bar{c}d} = \overline{\bar{a}b} \cdot \overline{\bar{c}d} = (a + \bar{b}) \cdot (c + \bar{d})$$

$$\overline{(\bar{a}b\bar{c} + d)e} = \overline{\bar{a}b\bar{c} + d} + \bar{e} = \overline{\bar{a}b\bar{c}} \cdot \bar{d} + \bar{e} = (a + \bar{b} + c)\bar{d} + \bar{e}$$

$$\begin{aligned} \overline{\bar{a}[c\bar{e} + b(\bar{d} + e)]} &= a + \overline{c\bar{e} + b(\bar{d} + e)} \\ &= a + \overline{c\bar{e}} \cdot \overline{b(\bar{d} + e)} \\ &= a + (\bar{c} + e) \cdot (\bar{b} + \overline{\bar{d} + e}) \\ &= a + (\bar{c} + e) \cdot (\bar{b} + d\bar{e}) \end{aligned}$$

The details of using the rules of complements have been stressed, so that the reader will gain a clear understanding, which should avoid errors particularly in the placement of parentheses.

6-4 LOGIC FUNCTIONS AND KARNAUGH MAPS

The illustrations of logic expressions by means of cells on a matrix graph have many advantages that will become evident as we proceed with our explana-

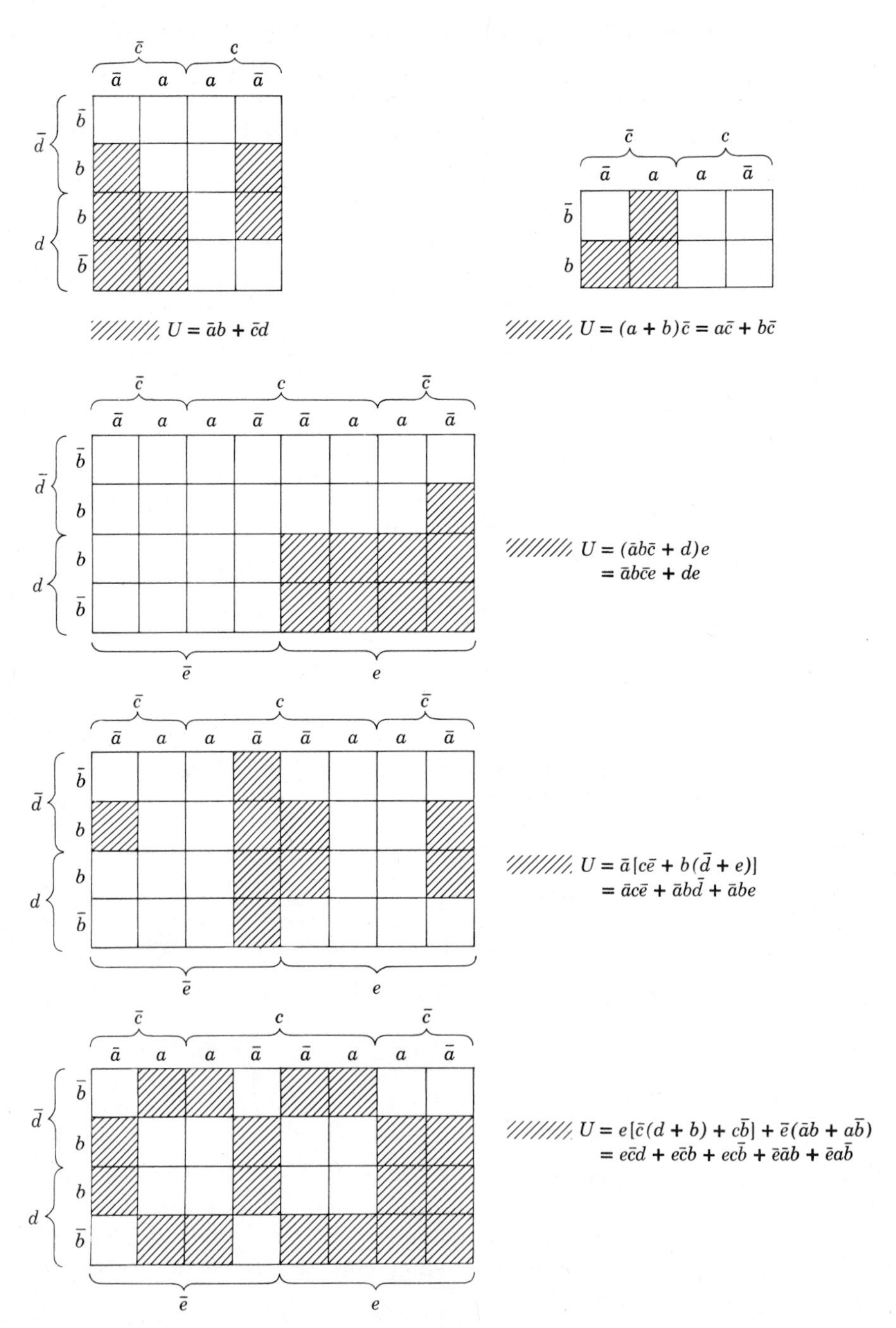

FIG. 6–9. Examples of logic functions illustrated by Karnaugh maps.

tion. This graphic process provides a method of comparing, transforming, and simplifying functions, just as we have already demonstrated by finding the complements in Fig. 6-8. Chapter 7 discusses a method of obtaining the required logic functions to solve automation control problems.

To express logic functions on Karnaugh maps, the following principles should be utilized:

- A function consisting of n distinct variables will require 2^n cells on a map.
- An AND function is the intersection of cells.
- An OR function is the reunion of cells.

Figure 6-9 illustrates the principles above, with a few examples. To facilitate the Karnaugh mapping process it is advisable to reduce each expression to its simplest form.

6-5 SIMPLIFICATION OF LOGIC FUNCTIONS

The logic functions that result from an analysis of an automation control problem are not always reduced to their simplest form. A reduction of the logic term to their basics will facilitate the understanding and solution to the problem.
There are two possible methods:

- The first is the Boolean algebra method, using the characteristic logic relationships explained in Section 6-3-1.
- The second employs the Karnaugh maps for a representation of the logic functions introduced in Section 6-4.

Figure 6-10 illustrates Karnaugh map simplifications of logic expressions.

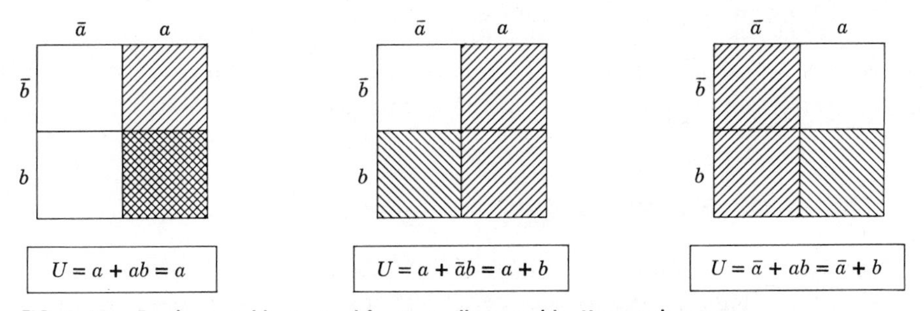

FIG. 6-10. Fundamental logic simplifications, illustrated by Karnaugh maps.

Some examples using the Boolean algebra method follow:

- $\boxed{U = (a + b)\ (a + c)}$ or

$$U = a \cdot a + a \cdot c + b \cdot a + b \cdot c$$

but

$$a \cdot a = a \qquad \text{(characteristic logic product)}$$

so

$$U = a + ac + ba + bc$$

$$U = a(1 + c + b) + bc$$

but

$$1 + c + b = 1 \qquad \text{(characteristic logic sum)}$$

so

$$U = a \cdot 1 + bc$$

$$a \cdot 1 = a \qquad \text{(characteristic logic product)}$$

Finally $\boxed{U = (a + b)\ (a + c) = a + bc}$

- $\boxed{U = \bar{a}b\bar{c} + \bar{a}c + a\bar{b} + \bar{b}.}$

Note.

$$ab + b = b$$

so

$$U = \bar{a}b\bar{c} + a\bar{c} + \bar{b}$$

but

$$\bar{a}b\bar{c} + \bar{b} = \bar{a}\bar{c} + \bar{b}$$

so

$$U = \bar{a}\bar{c} + \bar{b} + \bar{a}c$$

$$U = \bar{a}(\bar{c} + c) + \bar{b}$$

but

$$\bar{c} + c = 1 \qquad \text{(characteristic logic sum)}$$

Finally

$$\boxed{U = \bar{a}b\bar{c} + \bar{a}c + a\bar{b} + \bar{b} = \bar{a} + \bar{b}}$$

Figure 6-11 Graphically illustrates the examples above. The graphic method is frequently the easiest and fastest means of simplification.

During the process of graphic illustration, it is of value to express the $\overline{U}$ area on the map (the complement of the function). Once this is established, it leads directly to a simplified form of the U function.

6-6 THE UNIVERSAL LOGIC FUNCTIONS

As discussed in Chapter 2 a universal logic function provides the means to create all basic logic functions with combinations of identical devices.

The most popular universal functions are NOR, INHIBITION, and NAND.

The NOR Function (NOT-OR). This is the complement of an OR function

$$U = \overline{a + b}$$

Employing De Morgans' theorems the expression above can be transformed

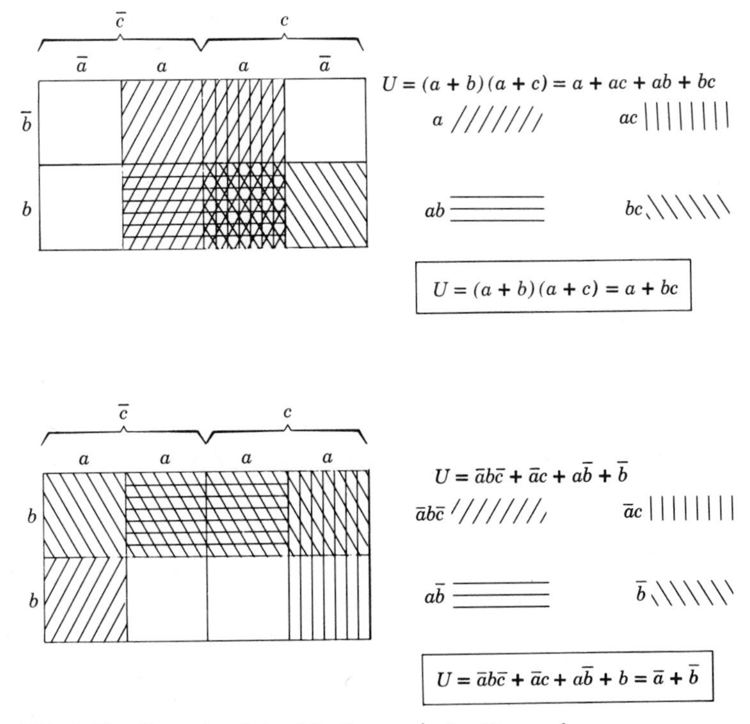

FIG. 6–11. Examples of simplification employing Karnaugh maps.

to $U = \bar{a} \cdot \bar{b}$. Chapter 2 describes how basic logic functions can be created with NOR functions or with INHIBITION functions (Fig. 2-4).

6-7 THE MEMORY FUNCTIONS

Chapter 2 discussed the memory function. Memories fill an important requirement in automation controls. When solving a particular problem, it is frequently difficult to perceive the requirement for the memory function. Chapter 7 discusses a method of facilitating the introduction of a memory into a circuit diagram.

A memory function is created by a binary device or circuit which retains the state (1 or 0) into which it was switched by the last control action which acted on the device, even if the action was only momentary.

In an automatic control circuit a memory serves two functions:

- It is a receiving device in that it receives signals which it must memorize; X_1 switches the memory into state 1 and X_0 switches it into state 0.
- It is a transmitting device in that it sends control signals to other devices in the circuit. Two output memories have the states x and $\bar{x}$ and one output memories have either x or $\bar{x}$.

CHAPTER 7

LOGIC CIRCUIT DESIGN METHODS

7-1 THE VARIOUS STEPS IN SOLVING AN AUTOMATIC CONTROL PROBLEM

Figure 7-1 shows some steps that help solve an automatic control problem with logic:

- The first step is to select the most suitable control hardware concept. This should create the best working circumstances and make the automation system more efficient as a result. This selection process includes everything, that is, power or output devices, and sensors as input devices both manual and automatic. It helps greatly to make this selection after one understands in detail the many fundamental forms discussed in Chapter 4.
- The second step is to find, that is, derive, the proper logic functions for the solution. These functions define the sequence of events that program the control circuit. Chapter 5 emphasized that fluid logic control is used mostly where machines are powered with cylinders. The method of "extended Karnaugh maps," especially useful here, is fully explained in this chapter and applied to a variety of problems in Chapter 8.
- The final step is to draw the circuit of the complete controller as defined by the logic functions. This is by far the easiest step.

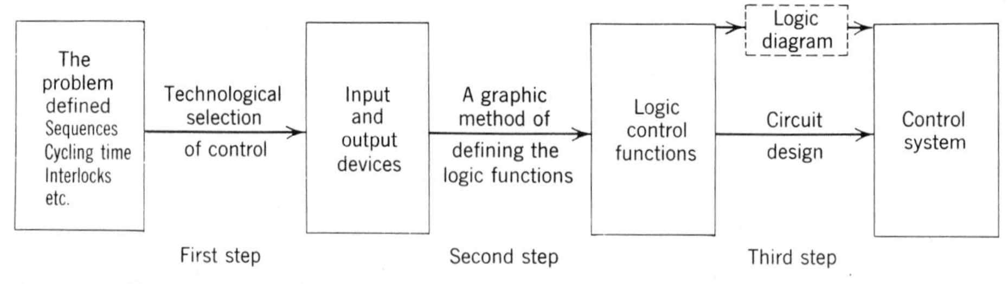

FIG. 7-1. The various steps that assist in solving automation control problems.

7-2 AUTOMATIC CYCLES IN GENERAL

The general principles to be outlined here are common to all automatic systems. To emphasize the more frequent situations, we discuss these principles in conjunction with their application to automatic control systems using cylinders to power machines.

The two cylinders shown in Fig. 7-2 are mechanically linked. They enable motion in two directions and are very commonly used on industrial machines to move objects from place to place during the manufacturing process. This movement is completely characterized by the four possible locations where the rod end of cylinder A stops after a motion is finished. Just where this rod end happens to be is detected by limit valves located at each of these four locations. They are the signal devices which program the controller and the next two examples show how useful they are.

7-2-1 The Square Cycle

Guided by Fig. 7-2, note that this circuit design is arranged so that each limit valve is actuated once per cycle. Each becomes the signal source for the next movement in the cycle, ordering actions on power valves which in turn actuate the cylinders, accomplishing the motion around the square pattern.

7-2-2 The L Cycle

Figure 7-2 also shows the fundamental difference between square and L patterns. The L cycle retraces steps and passes through point b twice per cycle. At point b, the movements ordered depend on whether the cycle is on its way to c or coming back from c. Order C_1 commands an action toward c, and C_0 reverses that movement. The limit valve at point b cannot differentiate these arrivals without help. Memory devices or circuits furnish this aid.

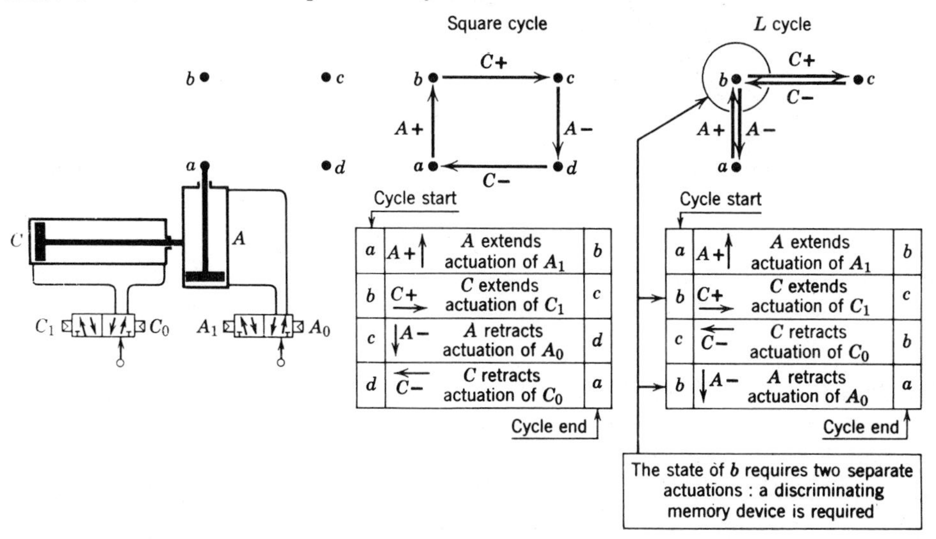

FIG. 7–2. Square cycle and L cycle with two mechanically interconnected cylinders.

This memory function is provided by the two states of a memory as discussed earlier. Assume that the memory function takes on state x when the rod end arrives at point c. Then bx is arranged to order A_0 as the cycle actuates the limit at b coming back from c. The next event in the cycle is actuation of limit valve at point a where the memory is switched to $\bar{x}$. Now the next arrival at point b sends the cycle on to point c because $b\bar{x}$ orders C_1. This use of the memory function has thus provided a positive method for interpreting the otherwise identical situations of actuating one limit valve at point b.

Perhaps a clearer illustration of this situation can be created by replacing the memory function with a third cylinder. Figure 7-3 shows a pattern of motion quite alike, but differs in that no point in space is passed twice per cycle. The cylinder end points could thus program successive steps around the pattern without aid of a memory, thus revealing the clue to a need for memory.

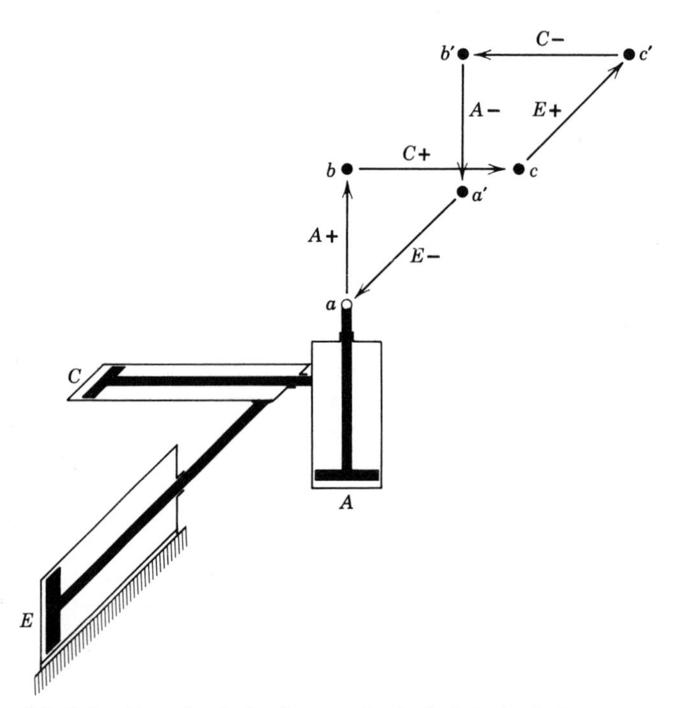

FIG. 7–3. Control cycle for three mechanically linked cylinders.

7-2-3 *L* Cycle Created with Independant Cylinders

Figure 7-4 shows yet another method for creating this pattern. Because they are independent of each other mechanically, this system requires two limit valves per cylinder to furnish sufficient information to program the proper motions.

The sequence of events listed indicates that state b AND c requires two different responses. Therefore, memory must be used to discriminate between the two. Since memory requirements become apparent during the develop-

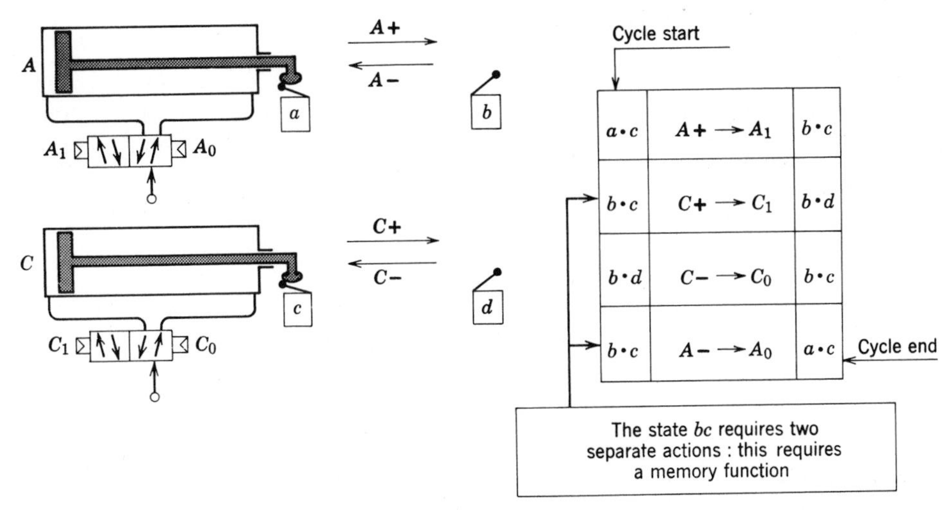

FIG. 7–4. *L* cycle with two independent cylinders.

ment and formation of primary variables from input devices to develop the patterns desired, they are known as secondary variables.

7-2-4 Combination and Sequential Cycles

By now you may recognize that the simple combination of variables at any step in the cycle can furnish the control signals required for the next action. Thus combination cycles only require primary variables.

Sequential cycles derive their control signals from present states and past actions. The past actions must therefore be stored by memory functions. Figure 7-5 illustrates a few typical examples of combination and sequential cycles. You can see that sequential cycles have at least one step in the cycle where the power path is retraced and the combination cycles do not.

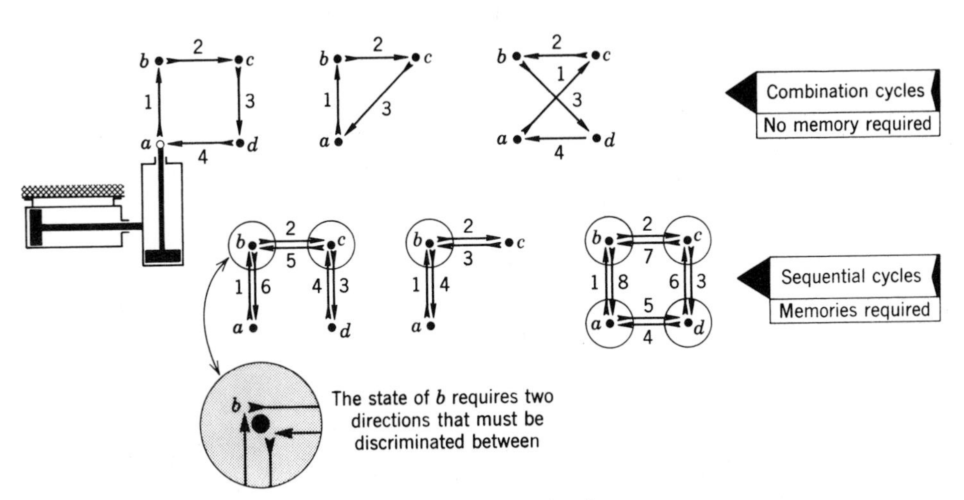

FIG. 7–5. Examples of combination and sequential control cycles.

7-3 EXTENDED KARNAUGH MAPPING METHOD

Maps are the graphic method for describing a problem. Karnaugh maps are used for developing the logic of solutions to industrial control problems. This method defines communication between variables and control signals and is most useful for introducing memory functions.

7-3-1 Basic Principles

Figure 7-6 shows the graphic steps for an *L* cycle employing independent cylinders. The map consists of a rectangular array of cells or areas. Each cell

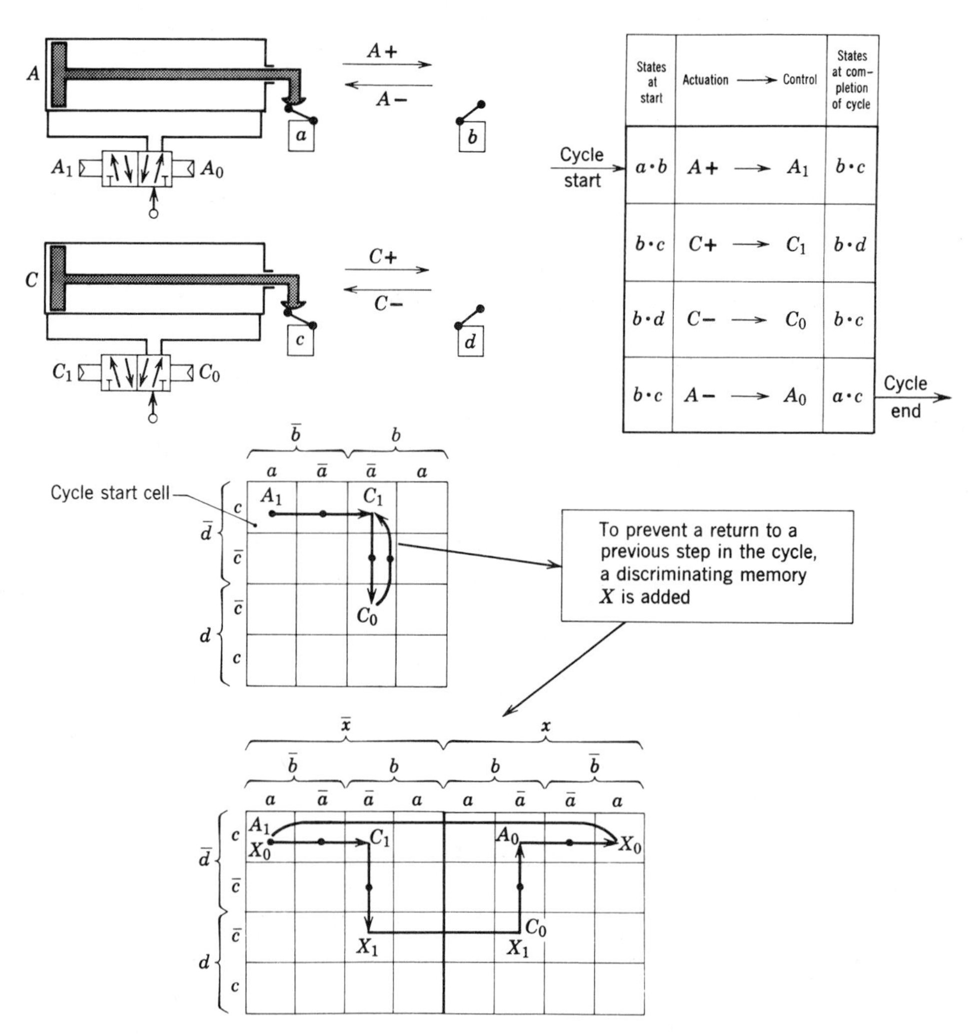

FIG. 7–6. Graphic illustration of the *L* cycle for two independent cylinders employing the "extended" Karnaugh mapping method (full size map).

identifies a particular combination of the primary variables a, b, c, and d. Let us start this cycle in the upper left corner cell which corresponds to states $a\bar{b}c\bar{d}$ of the cylinder locations.

The first action of the cycle is $A+$. This movement is ordered by control signal A_1 and to keep track of events on the map, A_1 is recorded in that cell, and for convenience hereafter, called the starting cell. The consequence of movement order A_1 is to extend the rod of cylinder A, releasing limit valve a. This produces a transient state $\bar{a}\bar{b}c\bar{d}$, and the dot in this cell is used to note the passage through this state. At the end of movement $A+$, limit valve b is actuated corresponding to the new states $\bar{a}bc\bar{d}$. The arrowhead in this cell indicates the completion of movement $A+$.

The second action, $C+$, is now commanded by order C_1; C_1 is recorded in the cell where it is issued. This order causes the rod of cylinder C to extend, releasing limit valve c when the motion begins and actuating limit valve d when the stoke ends. Cell $\bar{a}b\bar{c}d$ is occupied at this point in the cycle and corresponds to the end of movement $C+$.

Movement $C-$ is next on the program but order C_0 would simply cause step retracing; therefore, memory discrimination is necessary at this point. Memory is introduced to the map by duplicating the entire group of cells to the right. One set of cells corresponds to memory state x and the other set to memory state $\bar{x}$.

Now the control signal X_1 is introduced to move from cell $\bar{a}b\bar{c}d\bar{x}$ to cell $\bar{a}b\bar{c}dx$. To help ensure the required operation, X_1 is maintained in this cell as a precaution especially for memory functions created by holding circuits. Movement order C_0 is given at this point, causing action $C-$ which brings the sequence into cell $\bar{a}bc\bar{d}x$.

Ordering A_0 in this cell moves the sequence into cell $a\bar{b}c\bar{d}x$. This completes the cycle, but it will not restart until the starting cell is occupied again, initiated by the control signal X_0 to move from memory state x to memory state $\bar{x}$.

The record of this sequence is the Karnaugh map, a graph of the cycle listing all of the necessary states of the variables which helps to derive the logic functions that are the movement orders listed below.

$$A_1 = a\bar{b}c\bar{d}\bar{x}$$

$$A_0 = \bar{a}bc\bar{d}x$$

$$C_1 = \bar{a}bc\bar{d}\bar{x}$$

$$C_0 = \bar{a}b\bar{c}dx$$

$$X_1 = \bar{a}b\bar{c}d$$

$$X_0 = a\bar{b}c\bar{d}$$

If these logic functions seem unduly complex for the relatively simple task performed, it is because they indeed are more complex than necessary. This method of mapping also lends itself to simplification of the functions quite directly. The process of reducing their complexity is to encompass as much area on the map per function as possible. The following two rules are sufficient to demonstrate how this works.

- Combine cells not otherwise occupied during the cycle into the function. Because these empty cells were not filled in the development of the cycles functions, they obviously cannot be required. For example, logic function A_1 loses variables *c, b,* and *d* because extending its area into empty cells means that those variables take on both states rendering them meaningless; $A_1 = a\bar{x}$ is sufficient.
- Maintain the ordering signals as long as possible into the cycle. The orders for movements in this example are impulses that are effectively memorized by positions of the power valves. Control signal A_1 switched its power valve, whose position is maintained until A_0 signals its reversal. Therefore, A_1, the only other command for that cylinder, can be maintained up to the cell preceding the initiation of A_0. On the Karnaugh map, this means that the variable surviving the simplification of the first rule, variable *a,* also takes on both states and becomes useless. $A_1 = \bar{x}$ remains and is sufficient.

Note, however, that the area of A_0 cannot be extended into the domain of A_1, its complementing order.

By employing these two simple rules, the previous functions in complete form are reduced to those simplified equivalents listed below.

$$A_1 = \bar{x}$$

$$A_0 = cx$$

$$C_1 = b\bar{x}$$

$$C_0 = x$$

$$X_1 = d$$

$$X_0 = a$$

Active Variables. An active variable is one that gives the movement order. For example, on Fig. 7-6 the cell preceding the appearance of A_1 corresponds to the signal to switch from memory state x to $\bar{x}$. This signal, variable state *a,* is the function X_0. That switch event produces the appearance of memory state $\bar{x}$, which is the function A_1. On this basis, the active variables for the entire list of control signals is listed below.

$$A_1 \rightarrow \bar{x}$$

$$A_0 \rightarrow c$$

$$C_1 \rightarrow b$$

$$C_0 \rightarrow x$$

$$X_1 \rightarrow d$$

$$X_0 \rightarrow a$$

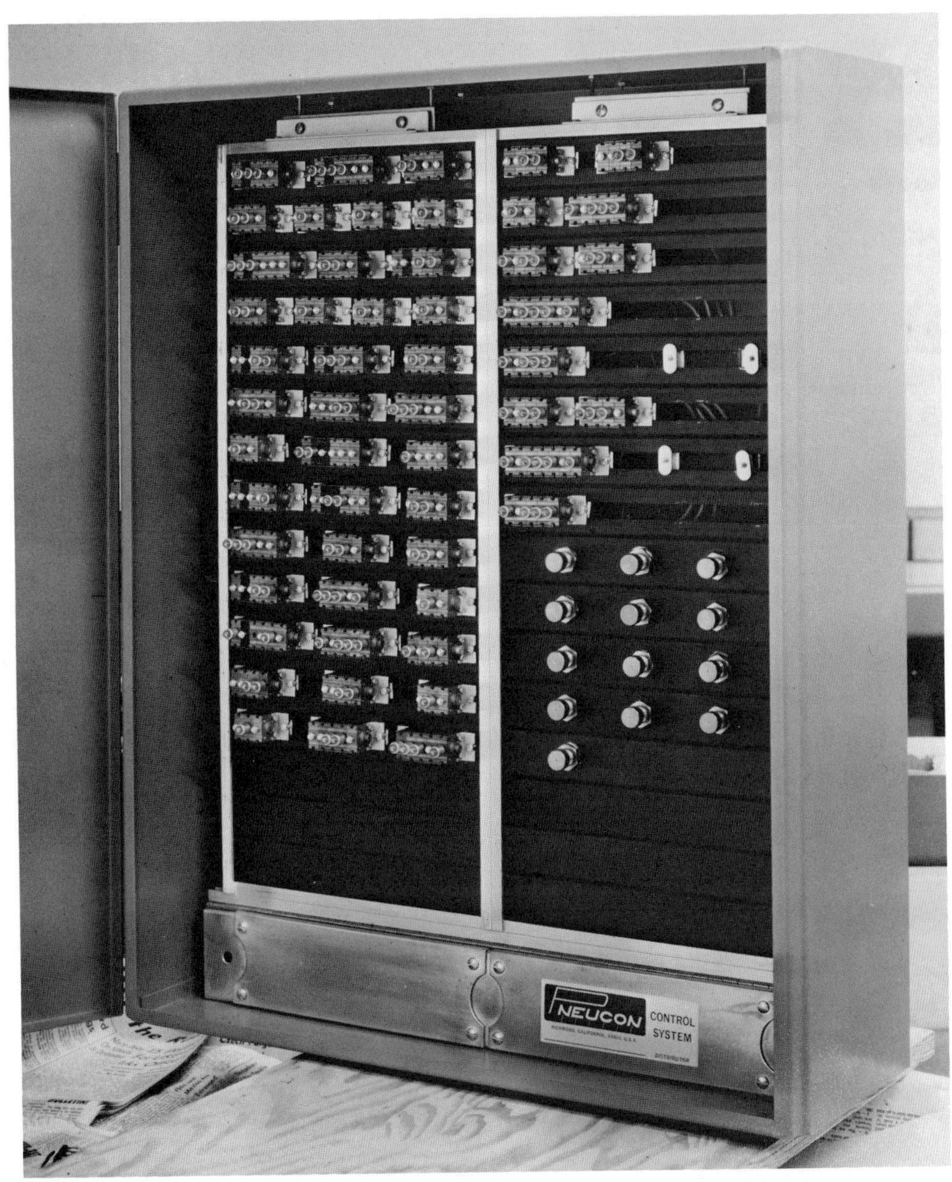

High-pressure movable-part system. (Courtesy Pneucon, USA.) The valve configurations shown here are constructed from individual body sections held together by clips. These sections, clipped together, reflect the size variety which make up standard valve and actuator types used to fit any circuit requirement. Note the pressure indicators plugged directly into the valve bodies. Manual controls are seen clustered in the lower right section of the cabinet.

This procedure makes the active variable quite easy to identify and helps simplify the logic functions directly from the Karnaugh map.

Incompatible Variables— Simplified Maps. Figure 7-7 illustrates two examples of how incompatible variables come about and how they are treated in the Karnaugh map. The limit valves along the strokes of the cylinders cannot be actuated simultaneously; therefore, the variables that they represent are indeed incompatible for they will never operate together.

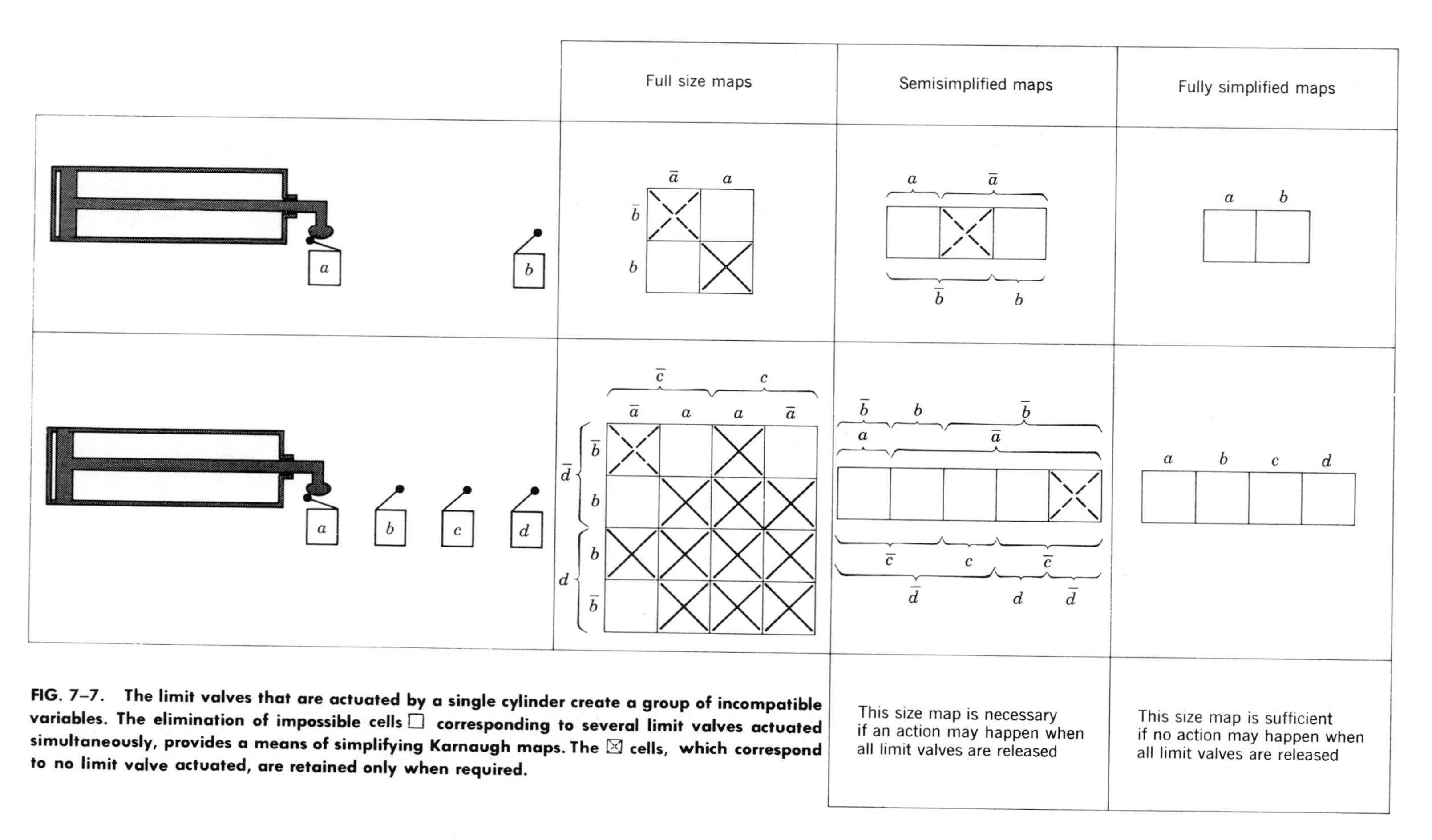

FIG. 7–7. The limit valves that are actuated by a single cylinder create a group of incompatible variables. The elimination of impossible cells ☐ corresponding to several limit valves actuated simultaneously, provides a means of simplifying Karnaugh maps. The ☒ cells, which correspond to no limit valve actuated, are retained only when required.

In practice, the development of automatic cycles produces many incompatible variables just this way. Since they never interact during the cycle, they can be dropped from the Karnaugh map and reduce the complexity of this task accordingly.

As an illustration, in the upper half of Fig. 7-7, variables a and b combined show the four possible states listed below.

- *a and b* actuated $\rightarrow$ ab $\rightarrow$ impossible therefore incompatible.
- a actuated and b released $\rightarrow$ $a\bar{b}$
- a released and b actuated $\rightarrow$ $\bar{a}b$
- a and b released $\rightarrow$ $\bar{a}\bar{b}$ $\rightarrow$ transient while cylinder moves.

These combinations in keeping with their significance produce the mapping circumstances listed.

- The complete map corresponds to the four possible states of variables a and b.
- The semi-simplified map retains the transient state $\bar{a}\bar{b}$, but has lost the incompatible state ab, and has only the three cells remaining.
- The simplified map contains the essentials for operation of the system, two cells corresponding to full travel of the cylinder, functions $a\bar{b}$ and $\bar{a}b$.

As a guideline, the following list of suggestions is helpful.

- Use complete maps when all variables are compatible.
- Use semi-simplified maps when the variables are incompatible, and a control signal is possible when all limit valves are released.
- Use simplified maps when variables are incompatible, and no control signal appears when all limit valves are released.

In practice, seldom is it necessary to initiate action when cylinders are in motion. The majority of our examples later on therefore, employ fully simplified maps. However, when such circumstances arise to produce the need for control signals during transient periods, the semi-simplified map is used as suggested.

The lower half of Fig. 7-7 illustrates this simplification procedure applied to a situation when four variables are incompatible.

Figure 7-8 illustrates the details for producing the L cycle with two independent cylinders by using a fully simplified map. The basic map is created by drawing a horizontal row of cells identified by the variables of cylinder A. To accommodate cylinder C, another identical set is constructed below them. Both rows are labeled on the side to produce the interaction possibilities of incompatible variables. As previously, the cycle originated in the upper left corner cell, state ac.

The cycle is drawn onto the map in accordance with principles outlined for the map in Fig. 7-6. The resulting logic functions developed are extracted

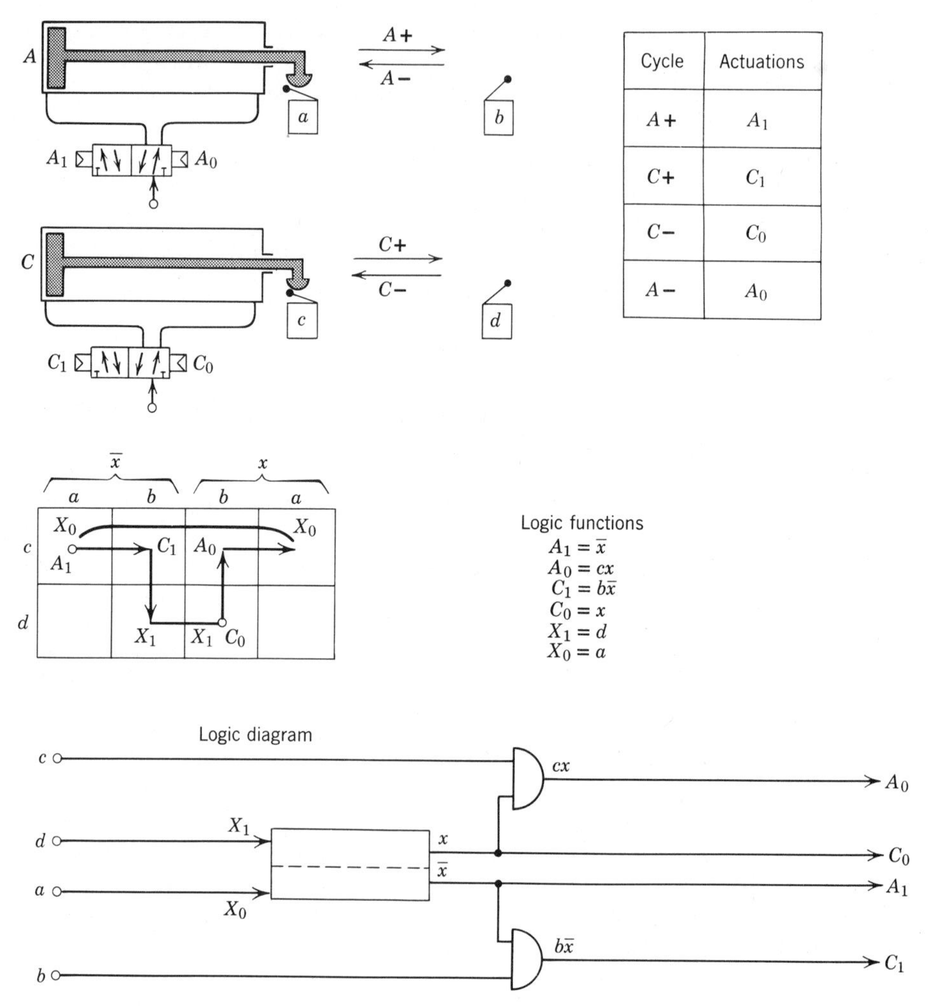

FIG. 7–8. Graphic illustration of the *L* cycle for two independently controlled cylinders, employing the "extended" Karnaugh mapping method (simplified map).

in their simplest form, the result of the simplified map, instead of the simplifying procedure used to produce them in the example of Fig. 7-6. Added on Fig. 7-8 is the logic circuit diagram necessary to program the cycle through the pattern.

Note in this example that the map is less complex, permitting the faster development of logic functions, a particular advantage when applied to circumstances where a large number of cylinders are necessary to accomplish a task. Large tasks, however, may produce situations where accidental actuation of limit valves could occur. In the foregoing example, protection against this possibility can be furnished by replacing b with $b\bar{a}$, a with $a\bar{b}$, c with $c\bar{d}$, and so forth wherever conflicts could develop.

Batch centrifuge controller. (Courtesy Fluidic Industries, USA.) This open cabinet typifies the hardware layout of an explosion proof controller used in chemical plants. Visible in the base section are filters, regulators, gauges, and a row of five output signal valves. Recognizable just above them are four adjustable timers on the left and five plug-in logic modules of the laminar/turbulent variety on the right side.

7-3-2 An Explanation of a Three Cylinder Cycle

The details of a three cylinder cycle are illustrated in Fig. 7-9. This example adds to our knowledge by introducing the three new circuit problems listed below.

- The addition of a start button for the cycle.
- The addition of a signal within the stroke limits of a cylinder.
- The addition of two simultaneous actions.

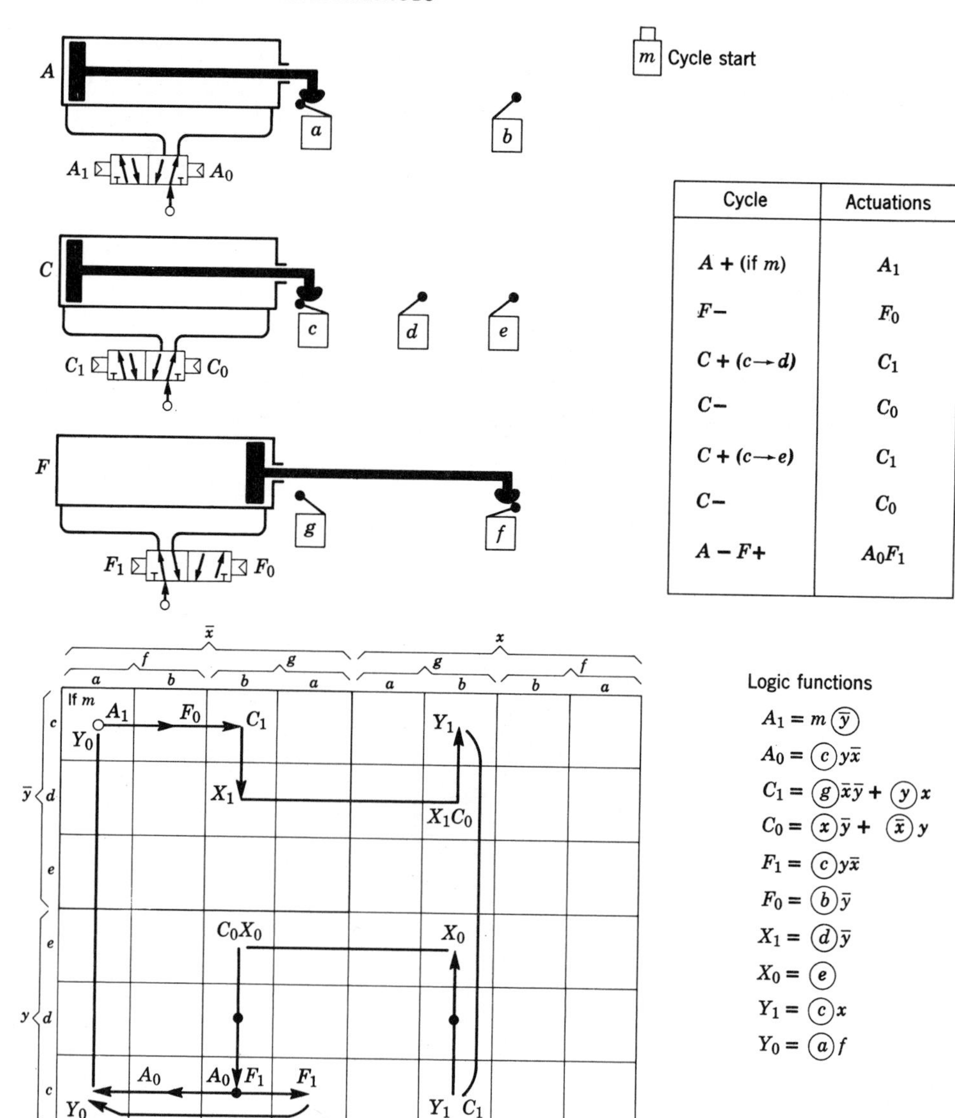

Cycle	Actuations
$A+$ (if m)	A_1
$F-$	F_0
$C+ (c \rightarrow d)$	C_1
$C-$	C_0
$C+ (c \rightarrow e)$	C_1
$C-$	C_0
$A-F+$	A_0F_1

FIG. 7–9. Graphic illustration of a three-cylinder control cycle.

7-3-2-1 Basic Map Construction in Fig. 7-9

Note the following:

- There are three groups of incompatible variables corresponding to the three cylinders.
- The description of the cycle defines that control signals are initiated only when a cylinder actuates a limit valve once the cycle has begun. This prerequisite facilitates using a simplified map.

The variable m, for starting the cycle, is not shown on the map. It is introduced after the logic functions are developed, as an external condition to complete the functions of this sequential cycle.

The basic map contains three rows of four cells aligned horizontally. The three rows correspond to the three cylinders, two having each a pair of interacting possibilities shown at the top of the map, all mapped to interact with cylinder C which controls three more variables. This basic group is outlined on the map and contains the starting cell at the upper left corner which helps the reading of the map.

7-3-2-2 *Drawing the Cycle in Fig. 7-9*

The first three movements are within the basic map, and to prevent retracing within this map on the fourth movement, $C-$, memory is introduced as outlined during earlier examples. Before entering another memory for the movement $C+$, a check to assure its need is useful. Here a retrace from memory state x to $\bar{x}$ would end in the cell already occupied with the command C_1. Thus the only solution is to add memory function y; this results in the whole map containing 48 cells.

When control signal C_1 is initiated for the second time movement c to e passes d without consequence, but a dot is placed in cell $bdgxy$ to identify this as an occupied cell. Then before cylinder C is retracted again, inversion of memory x is required. This time, during action $C-$, the actuation of limit valve c produces two actions. It is not possible to illustrate their priority on this simplified map, only the final positions of cylinders are recorded on the map, cell $bcg\bar{x}y$, where the cycle ends. In the event that both actions do not occur, the cycle stops running.

Finally, a return to memory state $\bar{y}$ completes the cycle by returning to the starting cell.

7-3-2-3 *Deducing the Logic Functions in Fig. 7-9*

To simplify the deduction of logic functions from this map, we use the method of active variables as discussed earlier in this chapter (Section 7-3-1). To begin, this example is limited to the memory area $\bar{y}$. Here the active variable for A_1 is $\bar{y}$ and is so entered in its logic function on Fig. 7-9. For reassurance that $\bar{y}$ is adequate, applying the first and second rules of simplifications discloses that some combinations can be discarded by using unoccupied cells, and those remaining through cell occupation are found unnecessary because there is no conflicting order for cylinder A in the entire area of $\bar{y}$. However, since A_1 is located in the starting cell, and a starting condition, m, has been introduced, the complete function is $A_1 = m\bar{y}$.

On the same basis, the active variable for A_0 is c but not all of the area defined by variable c can be used because some of it remains under the influence of A_1. Thus $A_0 = cy\bar{x}$ is the full retraction function for cylinder A.

The remaining functions are listed in Fig. 7-9 and the active variable for each is encircled for identification.

The + symbol is necessary to represent the functions C_1 and C_0. This is a result of the requirement for control signals to occur twice during the cycle at specific times and it is not possible to group them in the same area. The logic functions are the reunion of several areas.

Figure 7-10 illustrates the logic circuit for this example.

Some further simplification is possible by using a one-way limit valve

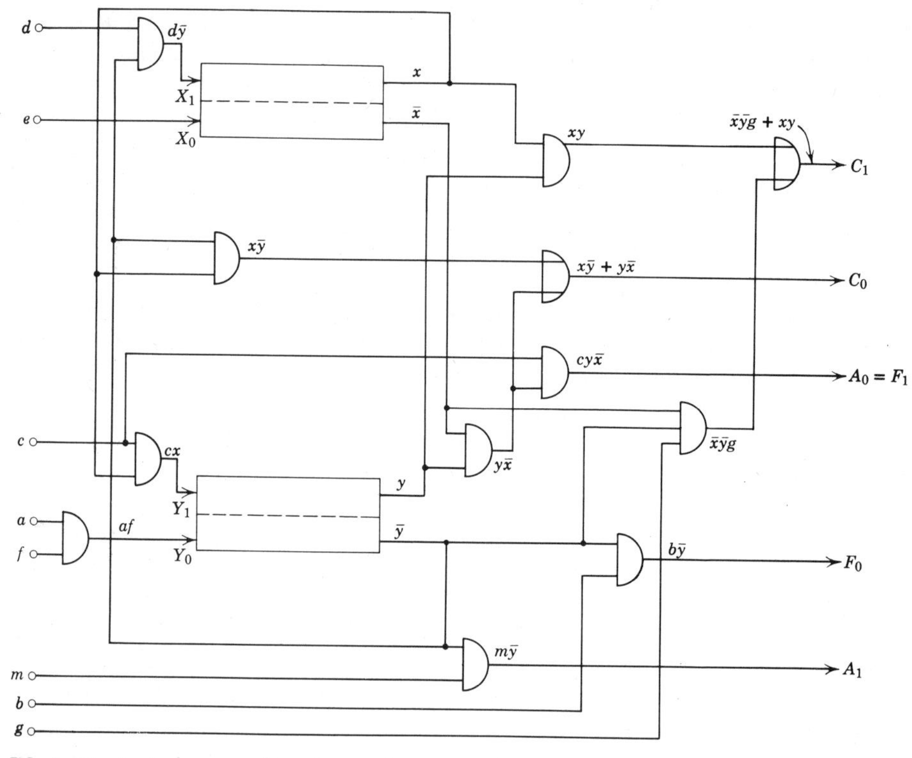

FIG. 7-10. Logic diagram of the three-cylinder control cycle.

at point d. If d is actuated when cylinder C extends, but not when it retracts, cell $bdg\bar{x}y$ remains empty. This enables the expansion of the influence of X_1 from $d\bar{y}$ to d, a simpler function. For this reason, intermediate signal locations usually employ one-way limit valves.

7-3-3 Using the Method

The method of extended Karnaugh mapping as just outlined can be put into practice with the help of four steps listed below.

- Define the incompatible variables.
- Draw the map.
- Draw the cycle on the map, remembering to add memory whenever the cycle returns to an occupied cell.
- Deduce the logic function by the application of active variables and the first and second rules to simplify functions.

The examples of industrial automation illustrated in Chapter 8 continue according to this list of steps.

7-4 FROM LOGIC FUNCTIONS TO CIRCUIT DESIGN

Although they were included in the illustration figures, so far the logic circuit or circuit design procedures have not been explained. The logic circuit is the implement most useful in constructing the interconnection of hardware. Its symbols were presented in Chapter 2. These symbols with input and output terminals, properly interconnected, constitute the logic diagram or circuit.

After the basic logic hardware form has been decided on, this circuit connection process follows the symbology related to that particular form as described in Chapter 3. Depending on the nature of the system, however, this procedure may not produce a controller with the minimum parts and interconnections so that still other simplification methods can be applied.

One of these is the grouping of common variables. In the example shown in Fig. 7-9, the combination $\bar{x}y$ is found in three functions, A_0, C_0, and F_1. There is no need to have three parallel circuit paths to perform this function in the circuit because just one of them will serve all three functions quite well.

The other is simply direct circuit analysis such as that depicted in Fig. 7-11, where an input function shows up in the same form two stages later in the connection diagram. That double transformation accomplished nothing

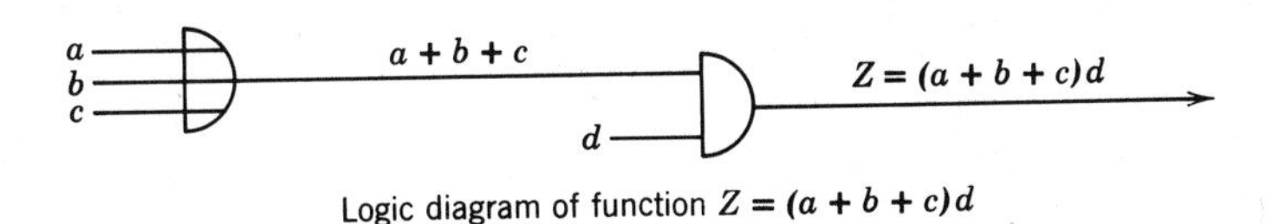

Logic diagram of function $Z = (a + b + c)d$

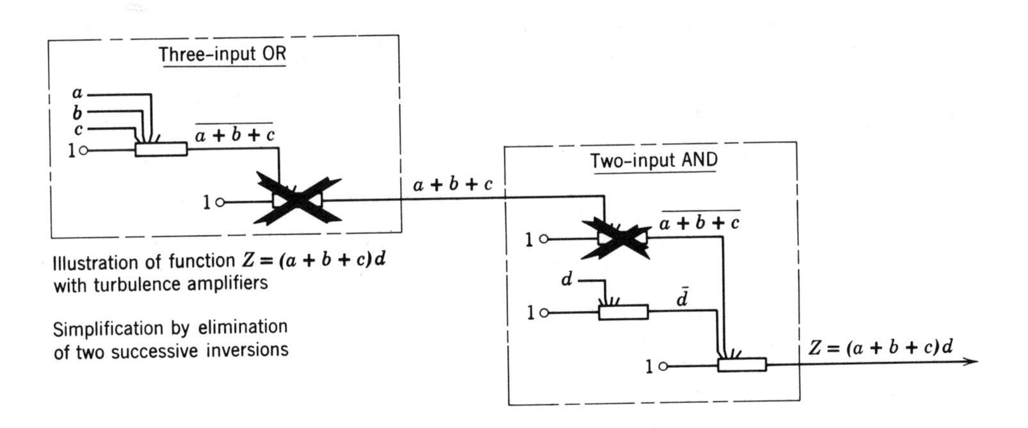

Illustration of function $Z = (a + b + c)d$ with turbulence amplifiers

Simplification by elimination of two successive inversions

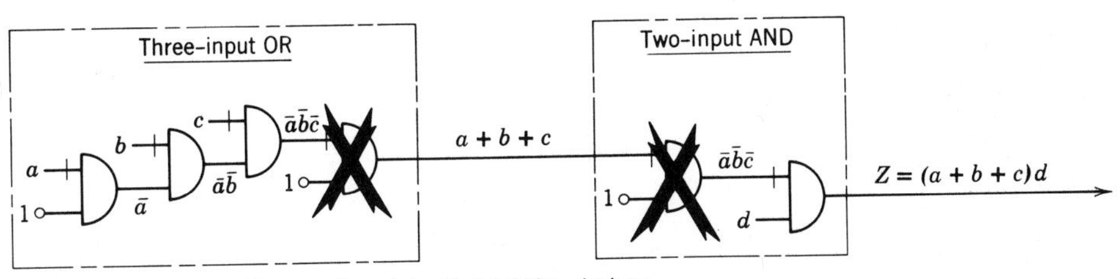

Illustration of function $Z = (a + b + c)d$ with inhibition devices

Simplification results from the elimination of two successive inversions

FIG. 7-11. Examples of circuit simplification.

and is therefore unnecessary in the control circuits shown and can be removed as suggested by the crosses.

7-5 LOGIC FUNCTIONS CREATED BY INPUT DEVICES

The majority of input devices, limit valves, sensors, and push buttons produce the YES function by initiating an output signal when they are operated. One exception is the interruptable jet which creates the NOT function; that is, the signal is not present when the jet is interrupted. This means that a judicious selection of the input devices for a control circuit can sometimes reduce complexity.

Still a larger variety of possibilities is provided by three-way valves commonly employed as input devices in conjunction with movable-part logic systems. Those types illustrated in Fig. 3-21, for example, can produce YES, NOT, OR, and AND functions simply by different connection procedures.

The utilization of this variety of circumstances can help minimize the number of logic devices when control circuits are designed but with the following disadvantages:

- More interconnections between the machine and the control circuit may result.
- Trouble-shooting may be more complicated.

Although this generalized procedure of transforming logic functions into logic circuits is useful for all types of systems, it is relatively common for particular manufacturers to suggest methods peculiar to their devices to minimize circuit complexity with their hardware.

CHAPTER 8

TYPICAL INDUSTRIAL AUTOMATION PROBLEMS SOLVED WITH FLUID LOGIC CIRCUITS

8-1 INTRODUCTION

In Chapter 7 we discussed the following topics:

- Logic function derivation.
- Logic function simplification.
- Logic circuit construction.

We now discuss the important subject of "technology selection." The typical examples illustrated and discussed in this chapter provide a practical basis for the important step of technology selection. The power system—cylinders, valves, and such—are mentioned in general. The controls, limit valves, and required peripheral equipment are discussed in detail.

Each typical application in this chapter illustrates our discussion in Chapter 5.

The simplified Karnaugh map method is applied to each example as soon as possible in the discussion and the examples are progressively more difficult; thus, the mapping method is easier to understand initially. The principles discussed in previous chapters are not reviewed; only new information is fully explained.

There are as many different approaches to discussing a control problem as there are possible solutions to a problem. Instead of illustrating the solution which is the most simple, reliable, or economical, the authors have endeavoured to illustrate the solution that is most suited in terms of "logical thinking." Excluding methods which are employed from habit, the logical method discussed here is the most easily applied to all problems and provides the reader with the means of solving all the problems that are typically encountered in industry.

8-2 SWITCHING CONTROL FOR THE SORTING OF CODED OBJECTS

This is a very common problem, frequently encountered in the sorting of various types of similar objects that are transferred on conveyor belts.

8-2-1 Problem Description

Figure 8-1 illustrates a solution to the sorting problem. Two cylinders mounted in tandem provide a means of shifting the conveyor in three discrete positions. The cylinders C and D are identical in bore and stroke sizes. We have the following:

- Position 1 $\rightarrow$ $C-$ and $D-$.
- Position 2 $\rightarrow$ $C-$ and D+.
- Position 3 $\rightarrow$ $C+$ and $D+$.

Cylinder A will stop the travel of the supply of objects during the time required for the routing of each object. Each cylinder is controlled by a four-way double piloted power valve.

The objects are coded by punched cards or combinations of cams. The code actuates two input sensors a and b in a combination which determines the direction of routing.

Input sensors c, d, and e are positioned: one on each of the routing paths to indicate the passage of an object and signal the release of the next one by the stop cylinder.

8-2-2 Operating Principles

When an object arrives on the conveyor, its code is read first by a and b ; then the following actions must be initiated simultaneously:

- Stop the arrival of the next object by extending the rod on cylinder A.
- Cylinder C and D are positioned according to the code read by a and b.

The object is routed in one of three directions. When it has passed the routing position, it actuates c, d, or e. This signal initiates the retraction of cylinder rod A. A new object is released for routing and the cycle repeats.

8-2-3 Variables

The two inputs a and b provide four possible combinations. We do not employ

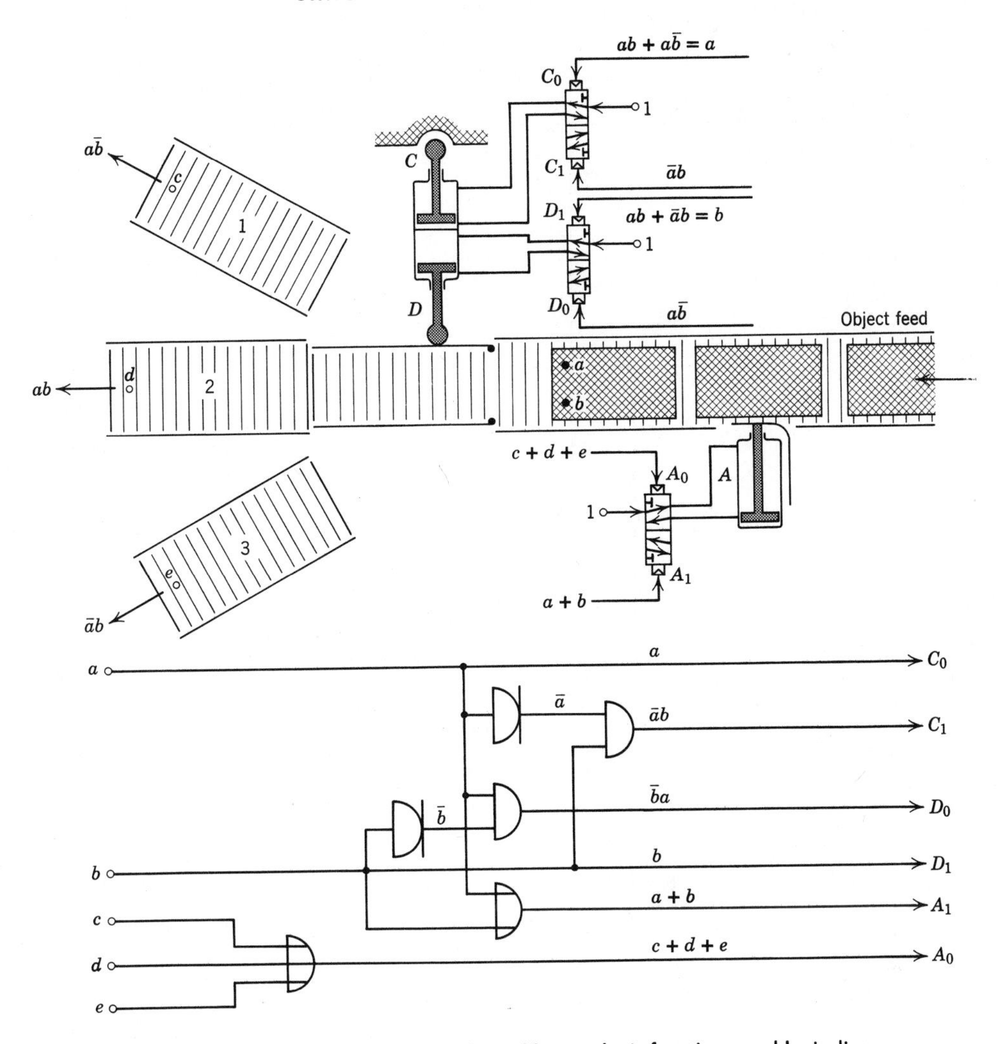

FIG. 8–1. Switching control of coded objects: physical layout, logic functions, and logic diagram.

state $\bar{a}\bar{b}$, which defines the absence of objects to be sensed. We arbitrarily assign the following values:

- State $a\bar{b}$ equals direction 1.
- State ab equals direction 2.
- State $\bar{a}b$ equals direction 3.

Logic Functions Defined. The problem is pure combinational logic. The logic functions can easily be deduced from the problem's functions:

$$C_0 = ab + a\bar{b} = a \qquad D_0 = a\bar{b}$$

$$C_1 = \bar{a}b \qquad A_0 = c + d + e$$

$$D_1 = ab + \bar{a}b = b \qquad A_1 = a + b$$

Figure 8-1 provides the logic diagram from the functions.

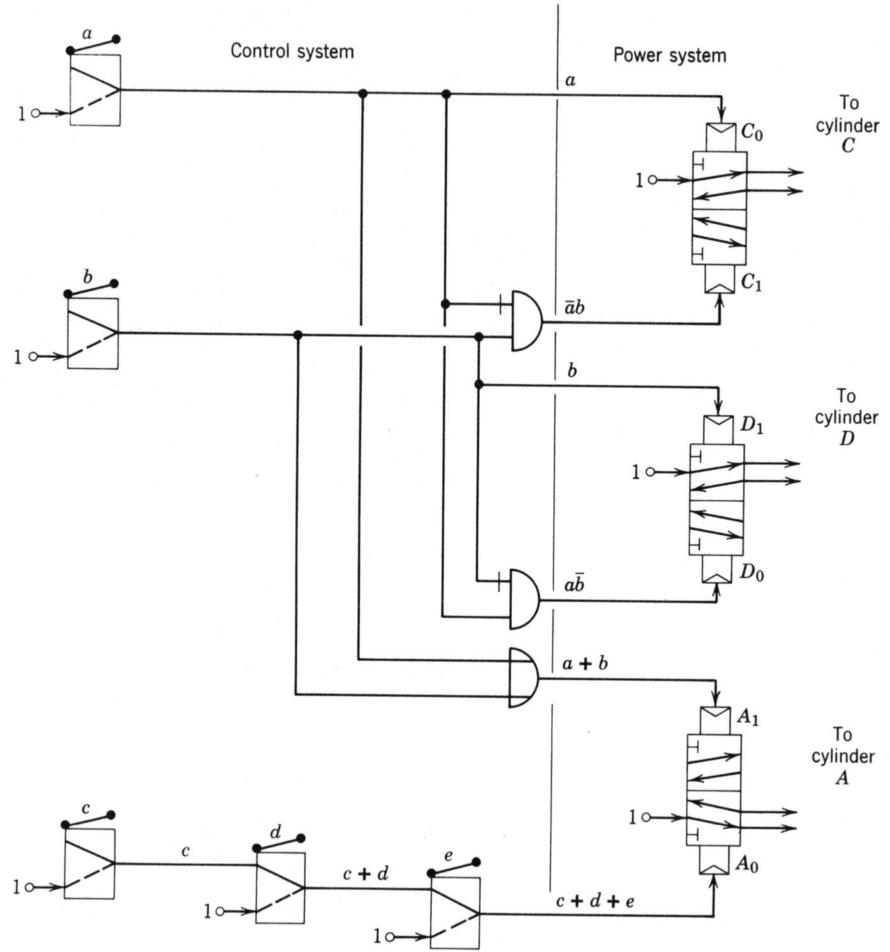

FIG. 8-2. Switching control for sorting coded objects, employing a high-pressure logic control system.

8-2-4 Selection of Technology and Circuitry

Two options are available; the selection depends on the objects to be sorted (large heavy objects or light small objects).

- When large heavy objects are to be sorted, mechanically actuated limit valves can be employed. In this instance, a high-pressure movable-part logic system would be practical, because the limit valves and the logic would operate on the same supply pressure; no interface amplifiers would be required. The circuit in Fig. 8-2 illustrates this type of logic system. Note that the A_0 function is created by the manner in which limit valves *c*, *d*, and *e* are connected to the logic.
- When small, light objects are to be sorted, nonmovable-part-type sensors (interruptible jets or back-pressure sensors) can be employed in conjunction with nonmovable-part logic systems. This type of circuit is illustrated in Figs. 8-3 and 8-4.

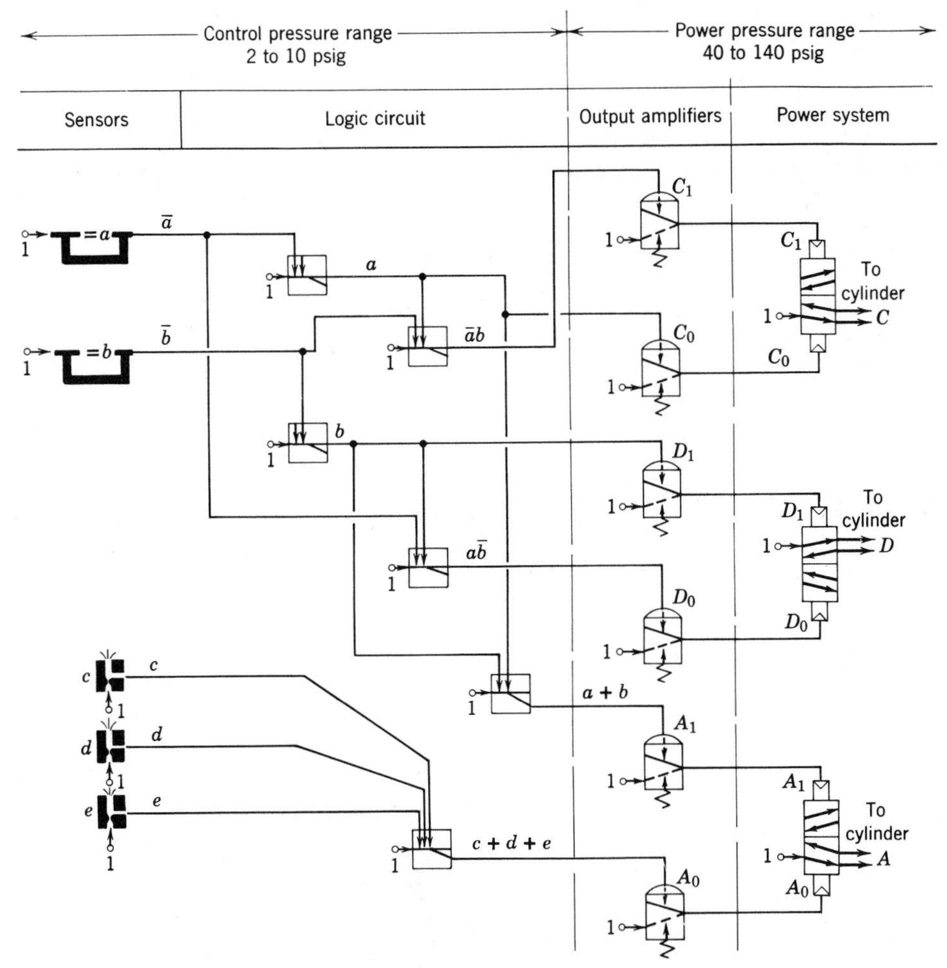

FIG. 8-3. Switching control for sorting coded objects, employing a jet deflection logic control system.

The circuit in Fig. 8-3 illustrates the circuit with the symbols for jet deflection logic devices. The circuit in Fig. 8-4 illustrates a circuit employing laminar/turbulent flow devices, which were explained in Chapter 3. Since these devices have a very low output level, the circuit has been designed to reduce the number of output amplifiers, which can be expensive in relation to the overall cost of the system. To help in the reduction of the cost of peripheral hardware, power valves with a single pilot have been selected so that only one output amplifier is required for each output.

The memory functions which are commonly created with double pilot power valves are in this instance created with logic devices (output memories).

Note. The circuits above are actuated when inputs a and b are sensed. When the objects travel beyond sensors, the transitory states $\bar{a}b$ or $a\bar{b}$ will be present before the states $\bar{a}\bar{b}$ are realized. The effect of these states can be avoided by the use of restrictors on pilot lines c and d.

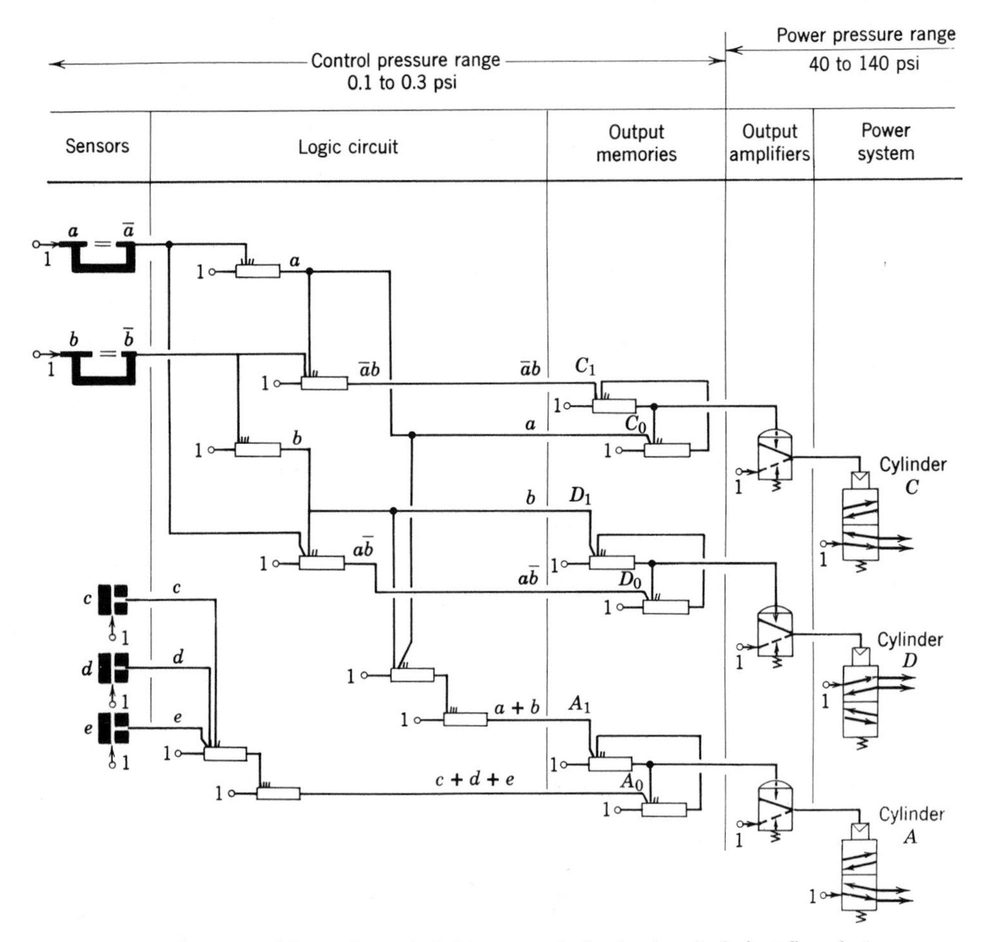

FIG. 8–4. Switching control for sorting coded objects, employing laminar/turbulent flow devices.

8-3 CIGARETTE CHECKING CONTROL

This problem is a good application for nonmovable-part sensors and fluidic-type Schmitt triggers.

At the end of a cigarette manufacturing line (production rate 10 cigarettes per second) each cigarette is checked for the density of the tobacco packing. An air jet is applied to one end of the cigarette; a sensor is positioned at the opposite end. The signal level for an acceptable cigarette must be within two extremes:

- A signal above the acceptable level indicates that the cigarette has been packed too loosely, because the air jet passed through it too easily. A Schmitt trigger A (see section 4-3-4-6) is set to switch above this level, providing logic signal a.
- A signal below the acceptable level indicates that the cigarette has been packed too tightly (the air jet flow has been restricted too

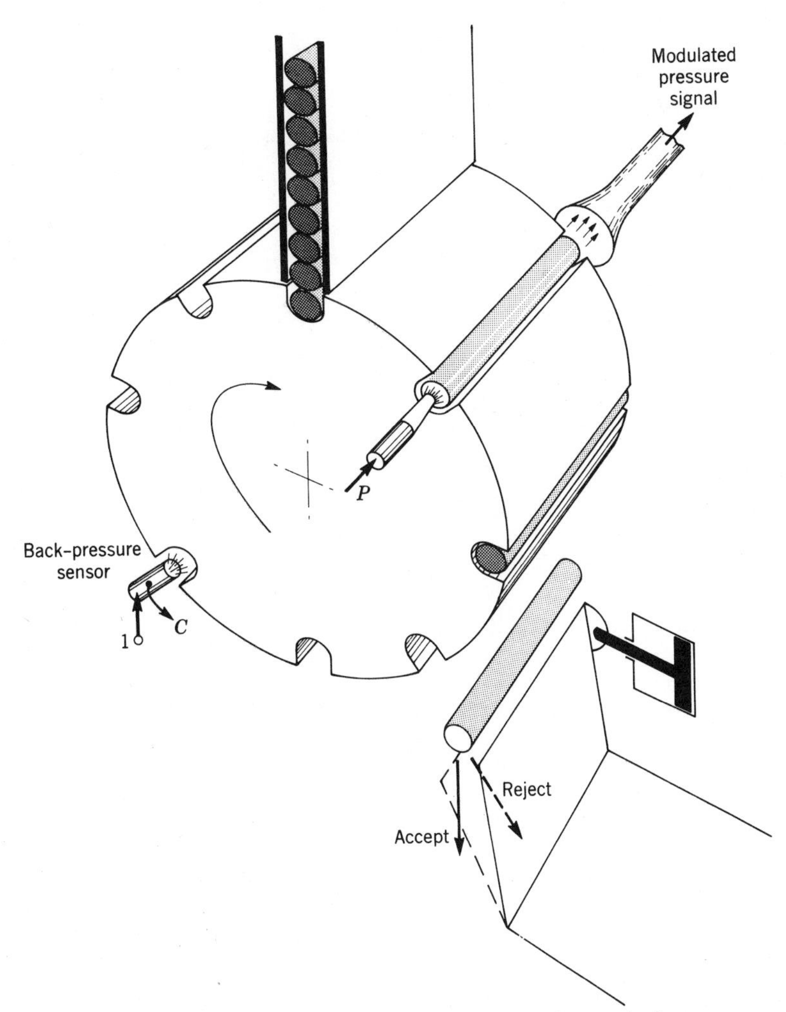

FIG. 8–5. Basic components of a typical cigarette inspection method.

much) or that the cigarette paper has a defect and allows the jet to leak to atmosphere. A Schmitt trigger B is set to switch below this level, providing logic signal b.

The cigarette is rejected if signal a or b is created. To provide this control at high production speeds, the cigarettes are transferred by a revolving drum with cigarette size slots (Fig. 8-5). A small air cylinder actuates a panel that channels the cigarettes as they drop from the drum to correspond to the accept and reject decision of the logic circuit.

As soon as the checking is completed, the sorting panel is positioned in the required position under the drum. Back-pressure sensor C creates signal $\bar{c}$ (no output when a slot is positioned in front of the sensor).

- When a cigarette is accepted, the retracting action equals $c\ (\overline{a + b})$.
- When the cigarette is rejected, the extending action equals $c\ (a + b)$.

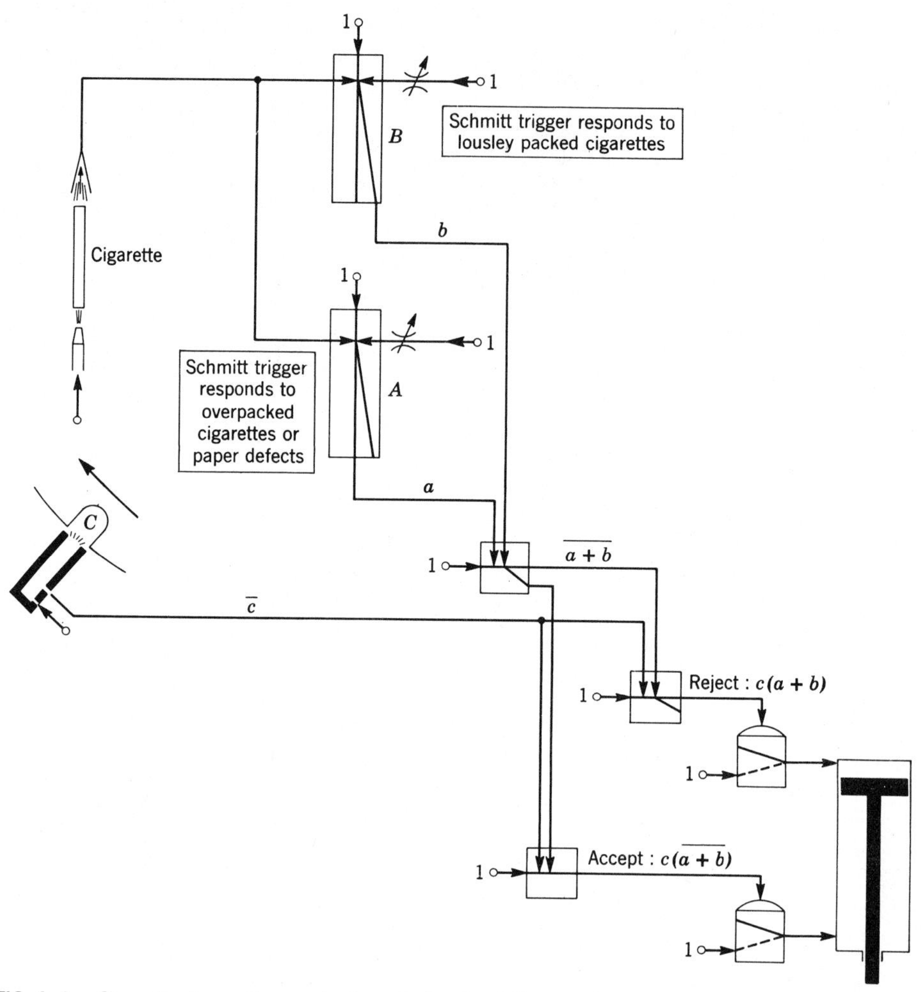

FIG. 8–6. Cigarettes inspection method employing jet deflection devices.

The high frequency rate creates a need for nonmovable-part logic devices. The circuit in Fig. 8-6 illustrates a simple and compatible system employing jet deflection devices.

8-4 THE SORTING OF OBJECTS AFTER DIMENSIONAL CHECKING

8-4-1 Description

The manufacturing of large quantities of a product creates control and sorting problems. If the manufacturing process is accomplished without machines, many people will be required and wide variation in the quality will result.

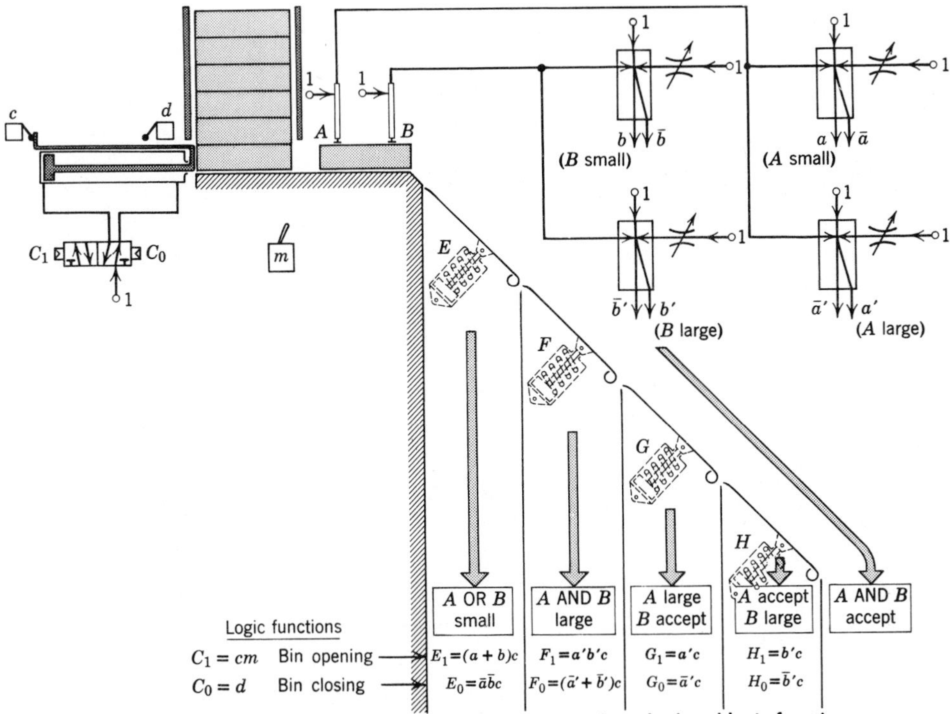

FIG. 8-7. Dimensional measurement and sorting by size: typical method and logic functions.

The automatic system illustrated in Fig. 8-7 can be applied in many instances, along with automatic dimensional checking and sorting of the products. Dimensions A and B are checked. Each of the two dimensional sensors provide a signal when the dimension is too large and another when it is too small. At the output the products are sorted into the five bins controlled by trap doors (Fig. 8-7).

Cylinder C feeds products from the magazine and positions them under the dimensional sensors. The feeder reciprocates continuously as long as the manual control m is "ON." If m is switched "OFF," cylinder C will always stop with its rod in the retracted position.

8-4-2 Selection of Technology

The selection is mostly of dimensional sensors. Section 4-3-4-6 and the previous problem have illustrated the usefulness of the Schmitt trigger. Figure 4-16 illustrates its application to dimensional checking.

This principle is employed in conjunction with a jet deflection logic system which provides a highly compatible system from sensors to outputs. Each of the two dimensional sensors A and B operates as two Schmitt triggers, one positioned for sensing of oversize and the other for sensing of undersize.

Signals a and b indicate that dimensions A and B are too small and signals a' and b' indicate that they are too large. Sensors c, d, and m are back-pressure sensors.

8-4-3 Definition and Circuitry

This is a pure combinational logic problem. The logic functions are provided in Fig. 8-7. The control cylinder C is functional and simple with its reciprocating action. When cylinder C rod is retracted the opening and closing of the trap doors is initiated; therefore, all functions depend on the state of C. Figure 8-8 illustrates the circuit converted from the logic functions. The fan out of four of some jet deflection logic devices necessitates the two devices for the function $\bar{c}$.

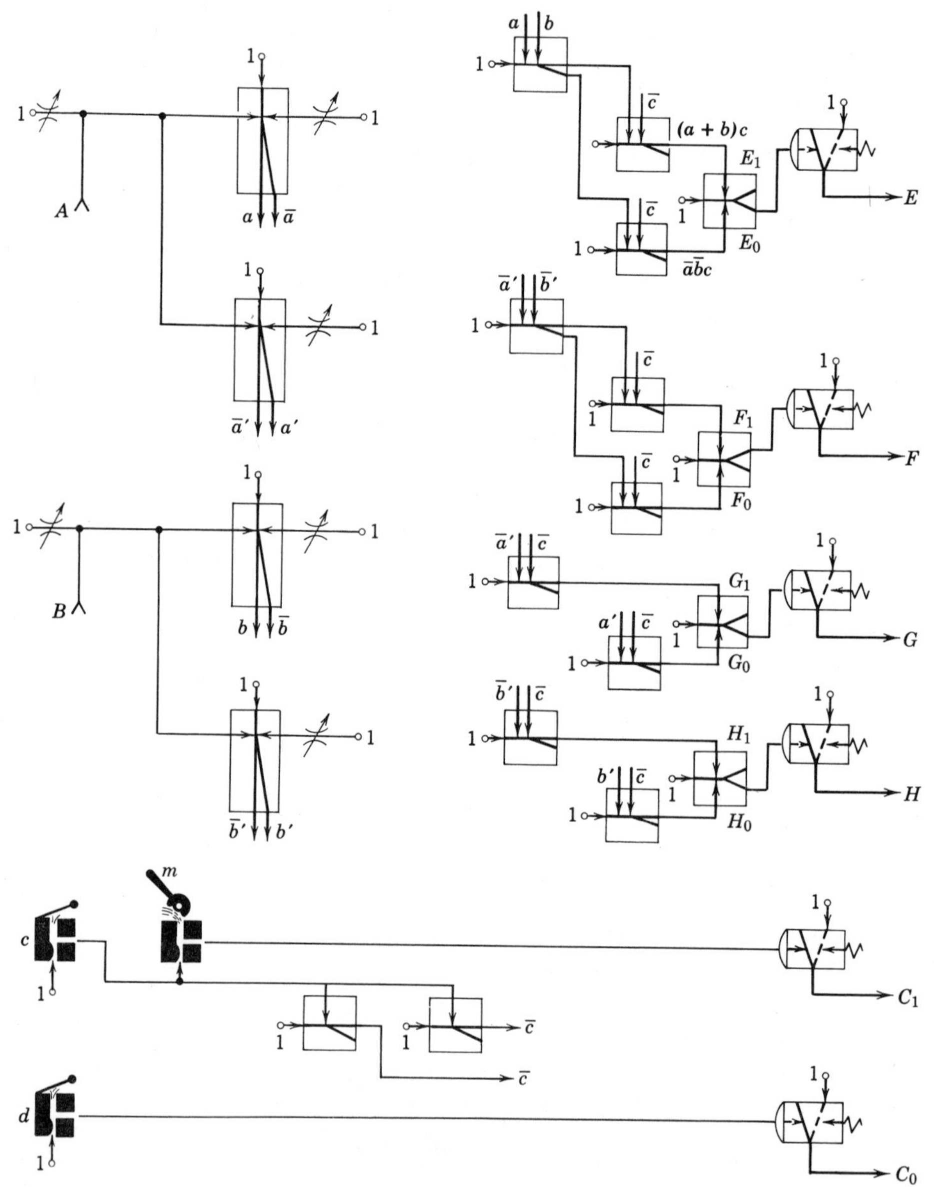

FIG. 8–8. Logic circuit for sorting objects after dimensional measurement, employing jet deflection devices.

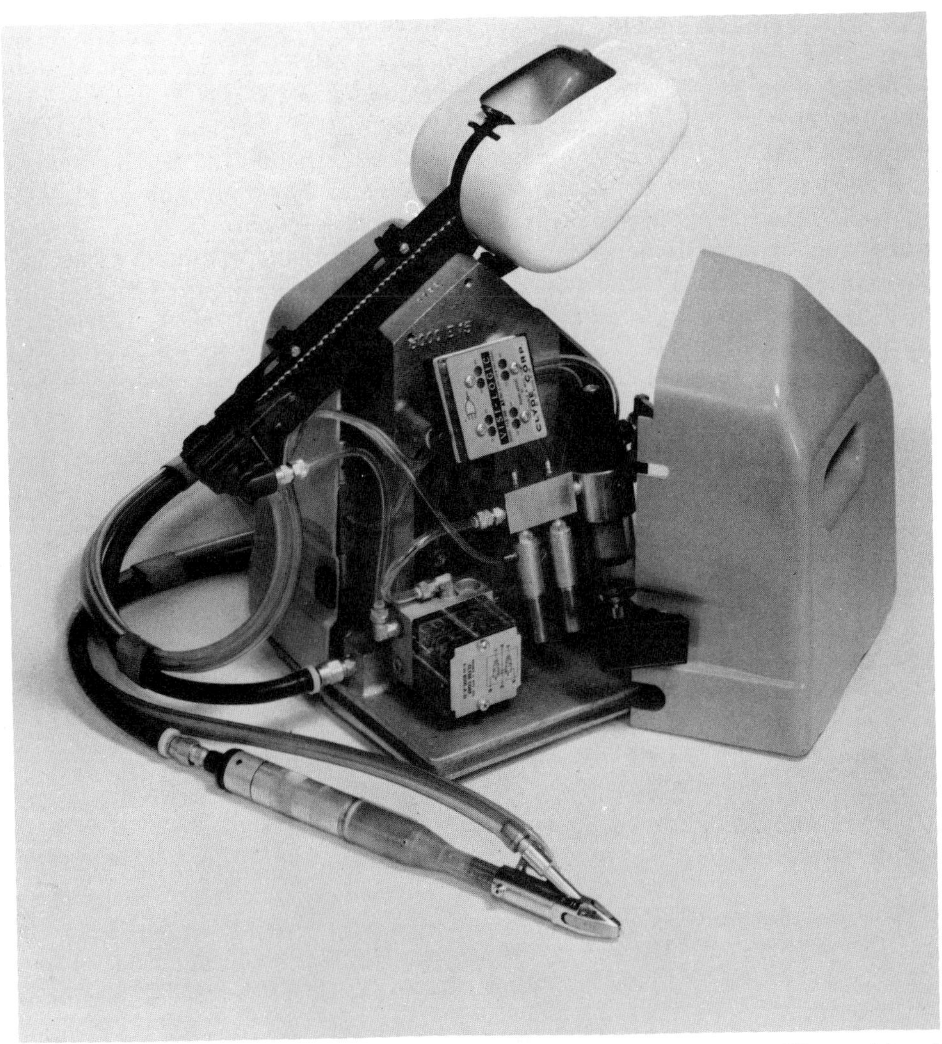

Pneumatic screw driver. (Courtesy Chicago Pneumatic Tool Co., New York, N.Y.) This portable unit illustrates an example of how air logic complements air power to make screw fastener driving automatic. Visibly identifiable components mounted on the machine tool are filters, regulators, and power valves with the logic module of the type described in Figure 3–23 located just above them.

8-5 TWO-HAND NO-TIE-DOWN CIRCUITS

When the operation of a machine is not fully automatic, the assistance of an operator is required. If the operator's hands must enter a dangerous area for loading or positioning of parts the physical hazard can be reduced by the use of start controls containing two-hands no-tie-down circuits. This type of start circuit requires the operator to depress two manually operated valves simultaneously to satisfy the start function, thereby protecting the operator's hands from injury.

The simplest possible type of two-hand circuit uses two manually operated valves connected in series. This circuit satisfies the basic two-hand actuation requirement; however, it allows the operator to circumvent the two-hand operation by tying down one of the manual valves. This process keeps one hand free, thereby negating the idea of the two-hand control. To eliminate the potential danger to the operator and to prevent the circumvention problem inherent in the circuit above, a "true" no-tie-down circuit must be used which requires simultaneous actuation of two manually operated valves and the release of both, before a new cycle can be initiated. Figure 8-9 illustrates two circuits employing high-pressure logic devices, which meet the requirements above.

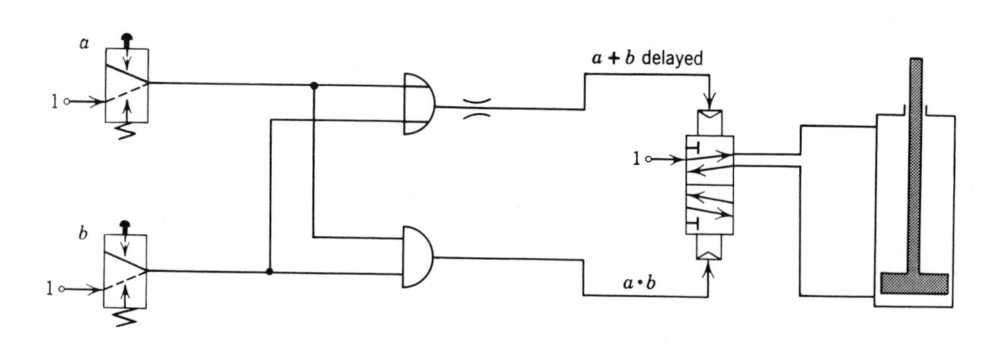

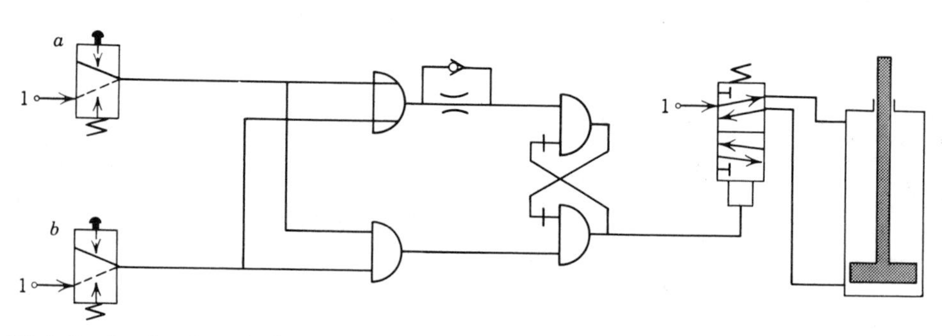

FIG. 8–9. Two hands no tie-down circuits.

The circuit on the upper half of the page is very simple. It needs only one AND and one OR device. If inputs *a* AND *b* are both actuated simultaneously, the outputs of the AND and OR devices will switch "ON." The OR output to the control valve will be delayed slightly because of the flow restrictor in the line. This delay allows the AND output to actuate the control valve, which causes the cylinder to extend.

When one or both of the manually operated valves are released, the AND output will exhaust faster than the OR output, because of the flow restrictor. This situation causes the delayed presence of the OR output to pilot the control valve back to its start position, thereby causing the cylinder to retract. If only one of the manually operated valves is depressed, the OR output will reach the control valve pilot after a slight delay.

No action will result because the valve is already in the cylinder retract position. If the second manually operated valve is depressed after the delay of the flow restrictor is completed, the AND output will switch "ON," but the control valve will not shift.

This circuit has one fault: If the air supply to the control and power devices is interrupted during the extending action of the cylinder, it will stop before the fully extended position is reached. When the supply air is restored, the cylinder will complete its extend stroke without the necessity of the operator actuating the two-hand control. This situation is a potential hazard to the operator's hands.

The circuit illustrated on the lower half of Fig. 8-9 solves the problem above. There are other variations on this circuit which satisfies the "true" no-tie-down circuit requirements which we have excluded because of their similarity. The circuits can also be implemented with low-pressure moving and nonmovable-part devices. Completely integrated circuits are available which require only the connection of inputs and outputs, as illustrated in Fig. 8-7.

8-6 OBJECT HANDLING

8-6-1 Description of Operating Principles

Figure 8-10 illustrates this system. Objects of various thicknesses must be held in position 1 and transferred to position 2. Three mechanically interconnected cylinders provide this action.

- Cylinder E holds the object.
- Cylinder C provides vertical travel.
- Cylinder A provides horizontal travel.

The cycle is started by actuation of push button "m" and proceeds as follows:

- Holding of object $\rightarrow E+$.
- Escalation $\rightarrow C-$.
- Transfer $\rightarrow A+$.
- Deescalation $\rightarrow C+$.
- Release $\rightarrow E-$.
- Escalation $\rightarrow C-$.
- Transfer return $\rightarrow A-$.
- Deescalation $\rightarrow C+$.

A and C create a U cycle.

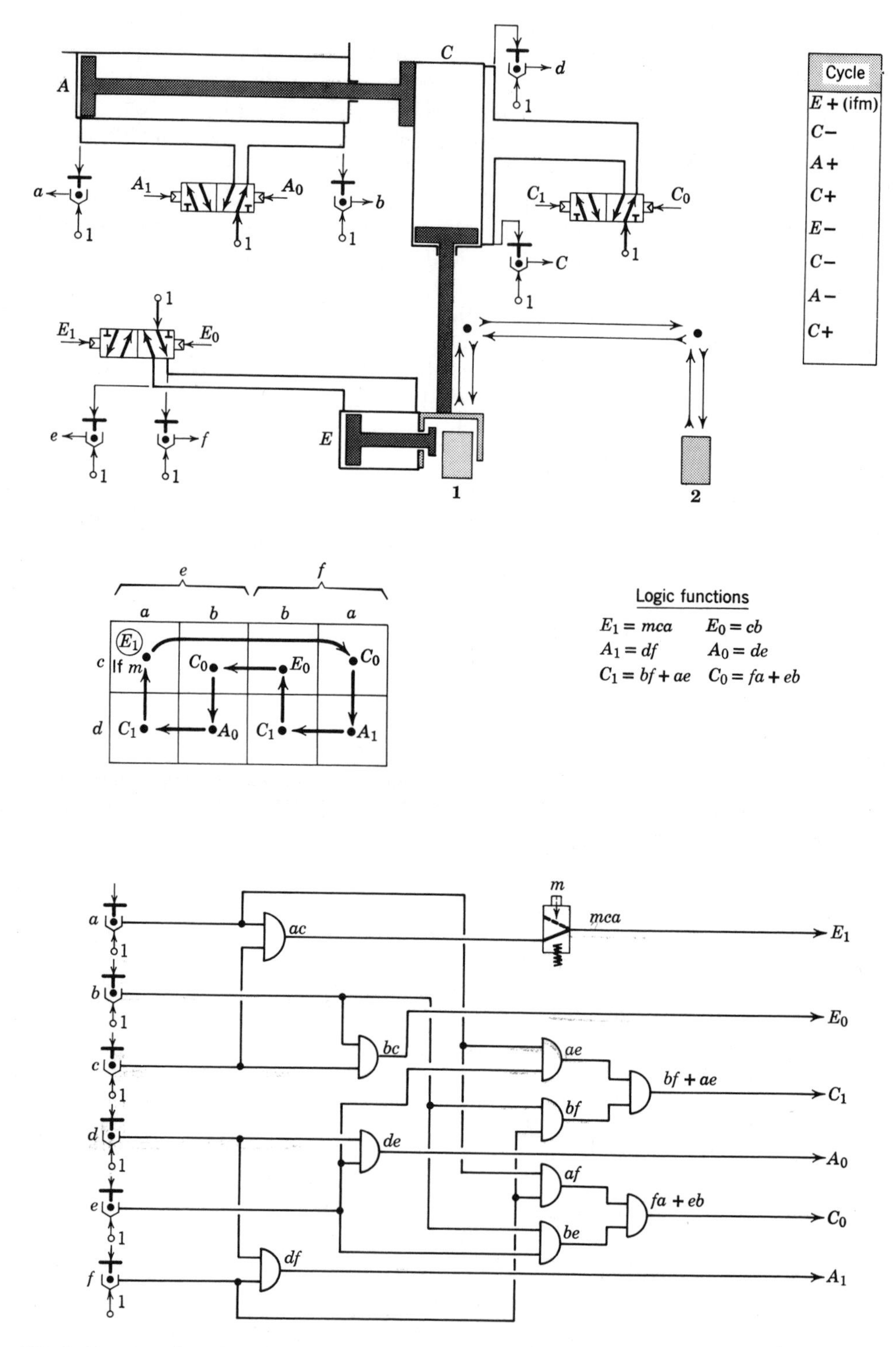

FIG. 8–10. Typical method of object handling. Cycle, Karnaugh map, and logic functions illustrated. Circuit employing high-pressure movable-part logic devices.

8-6-2 Selection of Technology

Pressure release limit valves are to be employed because of the following:

- The actions of the cylinders is not practical for the actuation of limit valves.
- The objects to be held vary in thickness.
- The positioning of the cylinder rods is not critical.

In connection with pressure release limit valves, a high-pressure movable-part logic system should be employed because of the same operating pressure requirements of the devices.

8-6-3 Definitions

The simplified Karnaugh mapping method provides a means of creating concise circuits. No action can be initiated while cylinders are in motion. We employ fully simplified maps. As illustrated in Fig. 7-5 a U cycle is a sequential circuit which requires a memory function. In this instance no memory function is required, because we are in a similar situation to the L cycle with the three cylinders described in Fig. 7-3. The third cylinder E, with its two states, differentiates between the two paths of travel in instances where the action in a cycle is repeated.

No particular difficulty is encountered in developing the logic functions. Figure 8-10 illustrates a circuit where all the components are compatible—they operate on the same pressure level.

8-7 PRESS AUTOMATION

8-7-1 Description and Operating Principles

Figure 8-11 illustrates a press consisting of two cylinders. This is a common type of press application in many industries.

- Cylinder C provides the pressure force.
- Cylinder A positions the object under cylinder C by positioning the table which has been loaded by the operator.

This type of press application creates a danger to the operator's hands. This problem is solved by creating a starting signal m with a no-tie-down circuit.

To ensure complete safety for the operator the actuation of the no-tie-down must be maintained during the first two actions of the cycle, $A-$ and $C+$ until the press cylinder is completely extended. If the operator releases the no-tie-down circuit during the two actions, the cylinders will return to the starting position. A time delay is initiated when the press cylinder is extended, to maintain this position for a specific time.

After the delay period is completed, regardless of the operators actions the cylinders return to the starting position in the following order: $C-$ and then $A+$.

To prevent an object from being pressed twice, an antirepeat feature is included in the circuit. At the end of the cycle if the operators action creating m is still present the next cycle must not repeat. To repeat the cycle the operator must release the no-tie-down circuit and then reactuate it again.

8-7-2 Selection of Technology

The majority of press applications employ hydraulics to create power for cylinder C. The power valve for this cylinder is a hydraulic type with pneumatic pilots. Mechanically actuated pneumatic limit valves c and d sense the ends of the rod stroke.

Cylinder A is a pneumatic type because its actions are fast and the required force is low. When the cylinder is in its retracted position under the press, the end of the rod stroke will be sensed with precision and safety by a mechanically actuated pneumatic limit valve b. At the opposite end of the stroke, a pressure release limit valve is employed.

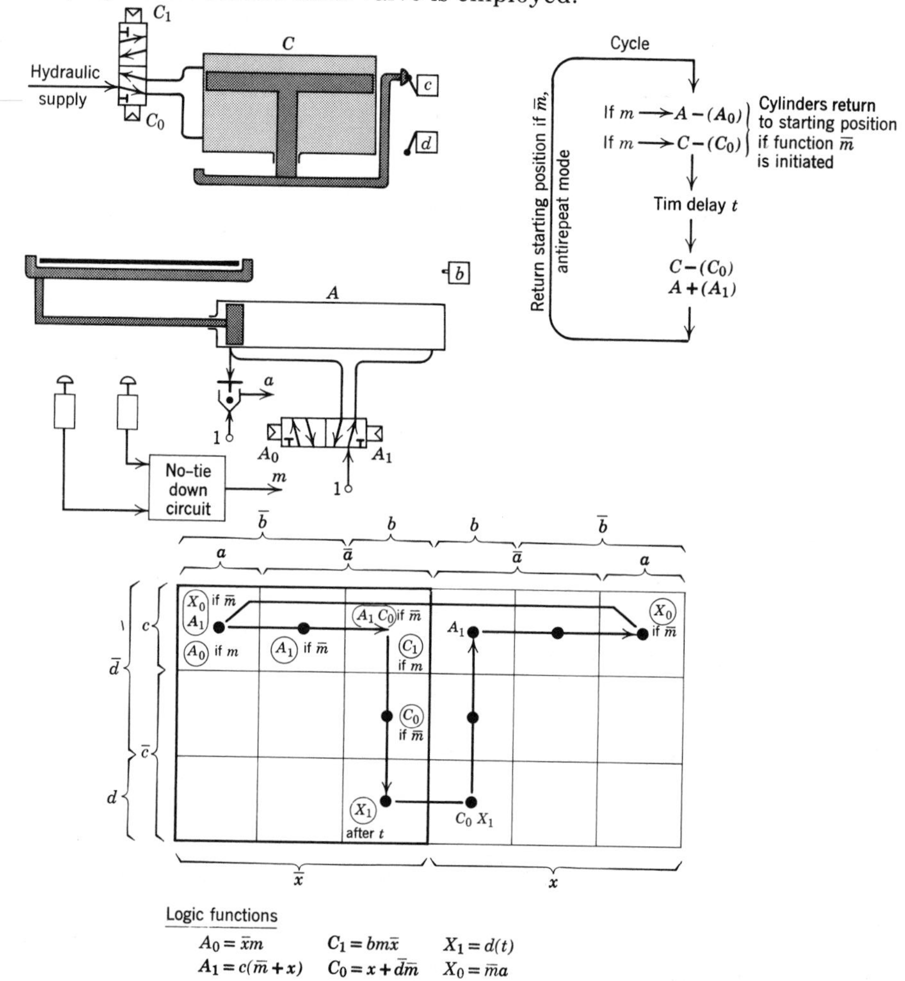

FIG. 8-11. Typical method of press control, Karnaugh map, and logic functions illustrated.

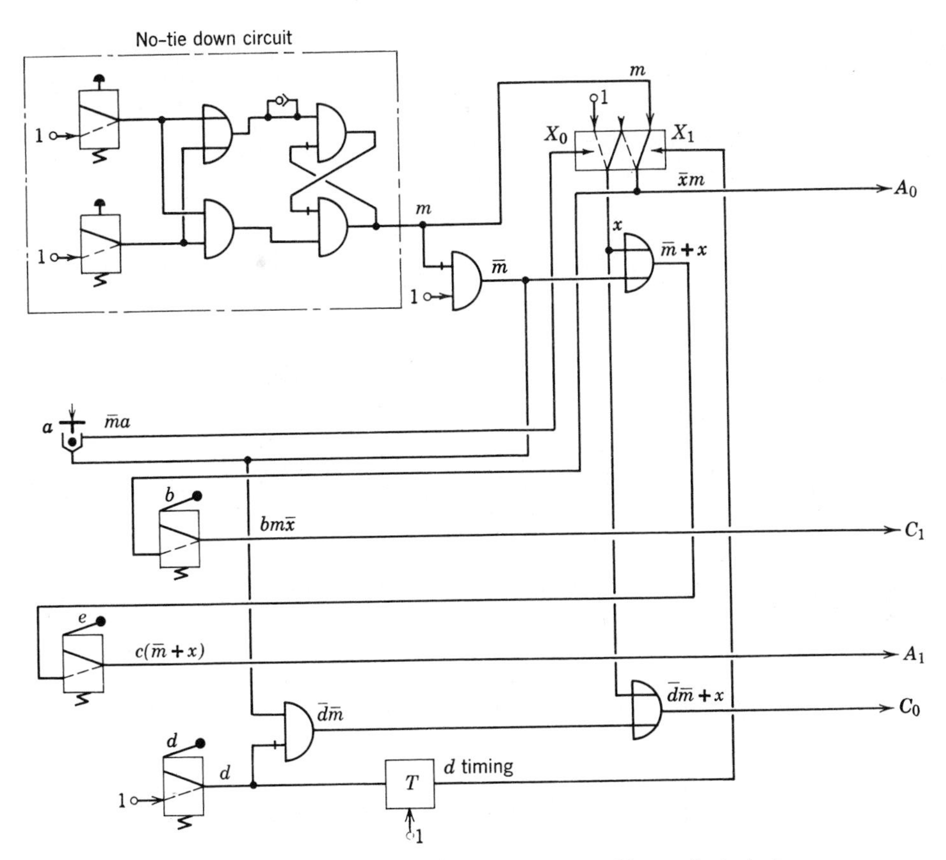

FIG. 8–12. Press control implemented with high-pressure movable-part logic devices.

8-7-3 Definition

A semi-simplified map must be employed in this instance, because m signal may disappear while the cylinders are in motion. If this occurs, an immediate counter action must be initiated. The Karnaugh map must illustrate the transitory states $\bar{c}\bar{d}$, and $\bar{a}\bar{b}$, between the rod stroke ends.

The cylinder return action illustrated on the Karnaugh map is created by subjecting the normal cycle to condition m. If $\bar{m}$ appears before the end of the press cylinder rod stroke, return action is initiated immediately. To solve the antirepeat problem, it is sufficient to introduce the $\bar{m}$ condition into the X_0 signal equation for the return to $\bar{x}$ state of memory X. If the operator has not released the manual controls, $\bar{m}$ is not present, so that a return to the starting cell is not possible and the next cycle can not be initiated.

Here we are providing a common method of integrating an antirepeat control into the circuit: the last action of the cycle does not end in the same cell if all the required conditions have been satisfied. The memory function is required if the circuit is to operate in the prescribed manner. Logic functions are easily derived from Karnaugh maps. Note that the functions in group $\bar{x}m$ are related to A_0 and C_1.

The circuit illustrated in Fig. 8-12 contains the following features:

- The two-hand no-tie-down function illustrated in Fig. 8-9.
- The unusual use of the spool valve memory X to obtain $\bar{x}m$ and x: the ports used normally as exhaust receive, in this application, one the signal and the other the supply.

Some logic functions are created by the manner in which the limit valves are connected in the circuit.

8-8 A SILK-SCREEN PRINTING MACHINE

8-8-1 Description and Operating Principles

This machine which is designed to print and cut labels from a tape is illustrated in Fig. 8-13. The power system consists of three pneumatically operated cylinders and a venturi vacuum pad.

- Cylinder A prints the pattern by spreading the ink across the silk screen.
- Cylinder C operates a cutting blade to cut the labels.
- Cylinder E moves the vacuum pad that holds the tape and transfers it through the sequence a step at a time.

A manually operated switch m controls the cycle as follows:

- Venturi is supplied with pressure. A vacuum pad holds the tape and after a short time delay, the rod retracts on cylinder E. The tape then advances one step forward.
- The cutter cylinder rod extends and cylinder A rod extends for printing.
- While the cutter is down the tape travel is stopped. Simultaneously the venturi supply is switched off. The vacuum pad releases the tape, and after a short time delay, cylinder E rod extends.
- The cutter cylinder retracts to the starting position.

The sequence of events above produces one label. At this step, in the sequence, cylinder A is not in the retracted position. The printing of the second label will retract cylinder A. The sequence for the second cycle is similar to the first, except that cylinder A action is reversed. The complete cycle of cylinder A will result in the printing of two labels.

8-8-2 The Selection of Technologies

Since the cylinders carry very little load, any incorrect operation would not result in any damage to the machine. A cam programmer, as described in Section 4-6, provides an efficient control. The employment of a cam programmer facilitates the tubing connections and sequence adjustments.

Each cylinder is controlled by one cam that actuates a mechanically operated pneumatic valve. No limit valves are required. This type of control system is a combinational type.

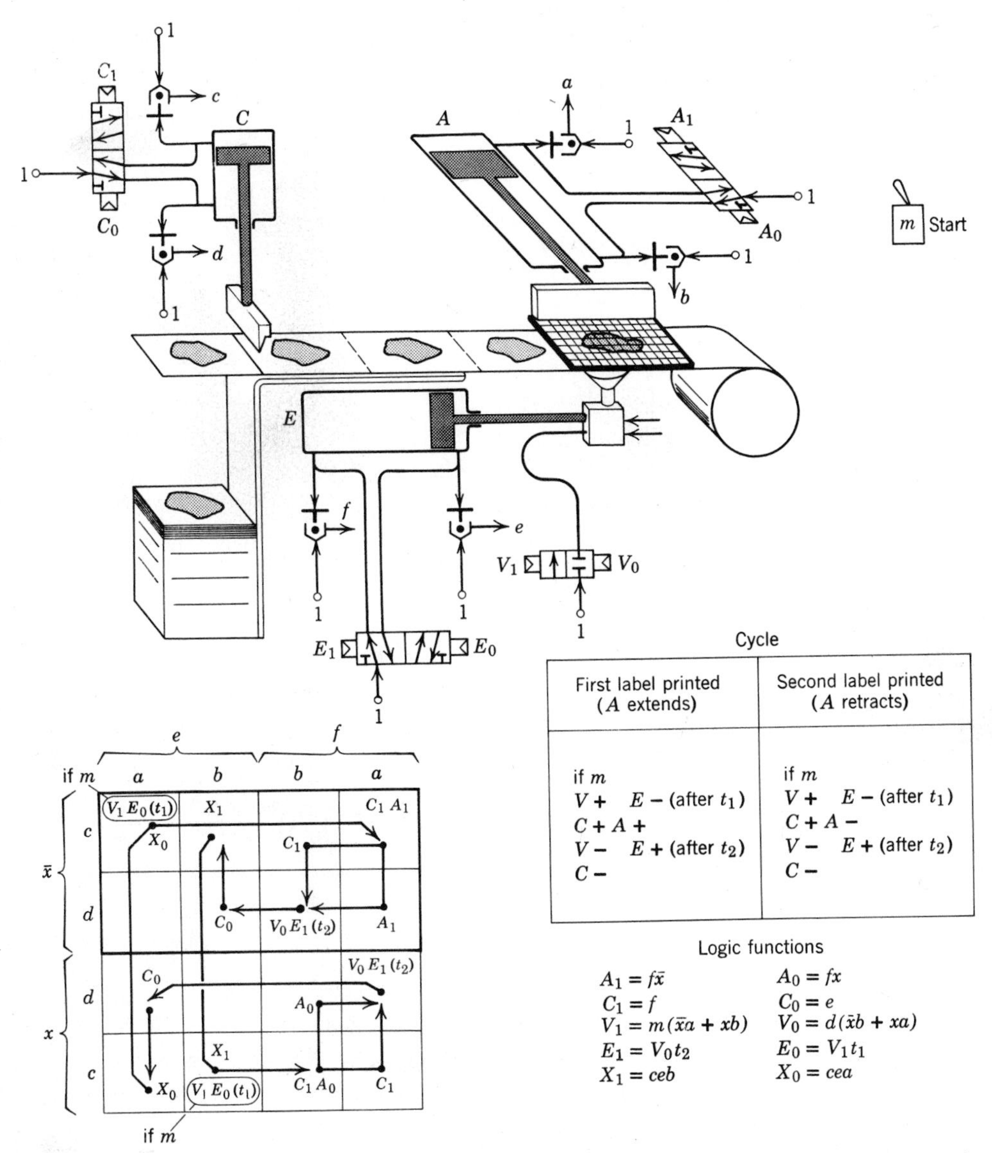

FIG. 8–13. Silk screen printing machine: typical design method, cycle, sequence, Karnaugh map, and logic functions.

A more reliable version of the machine above would be possible if a sequential control circuit with pressure release limit valves was employed instead of the standard type of limit valves.

8-8-3 Definitions

The cycle requirements are well suited to a fully simplified Karnaugh map, which can easily solve the problems. The two halves of the cycle are divided by the two states of the memory function.

The circuit illustrated in Fig. 8-14 employs a high-pressure air supply because of the operating requirements of the pressure release limit valves. Here again some logic functions are satisfied by the input devices.

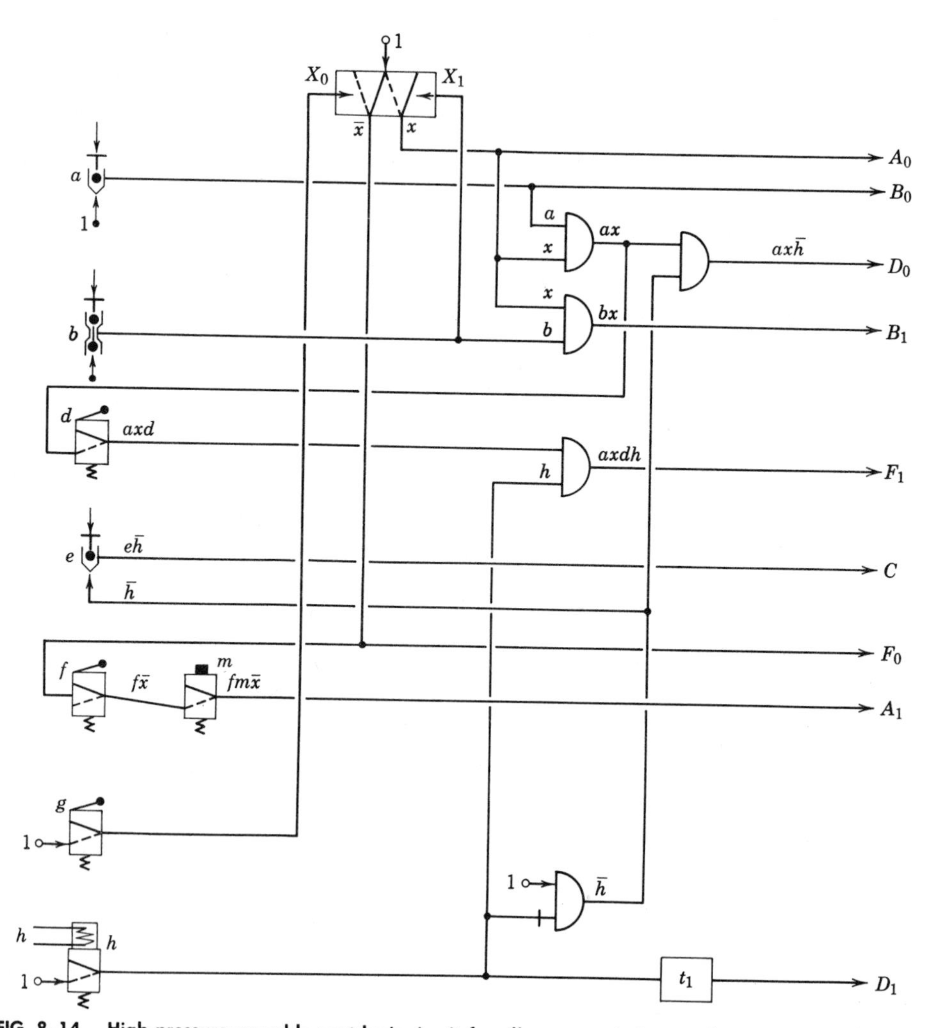

FIG. 8-14. High-pressure movable-part logic circuit for silk screen printing machine.

8-9 AUTOMATIC CONTAINER FILLING

8-9-1 Description and Operating Principles (Fig. 8-15)

Containers are filled automatically with a liquid until the weight reaches a specific level. This problem is encountered in the manufacturing of many products.

Containers are fed by cylinder C which pushes them one at a time on to

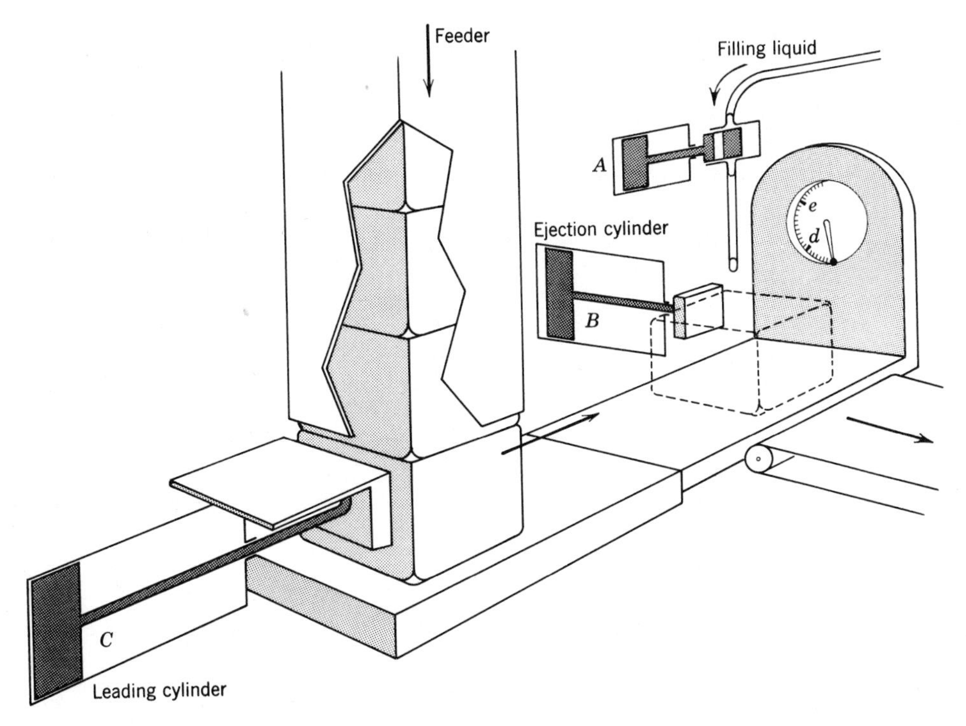

FIG. 8–15. Container filling typical method.

the scale for weighing. The weight of the containers on the scale actuates sensor *d* which causes cylinder *C* to retract and valve *A* to open and supply the liquid to fill the container.

As soon as the volume of liquid in the container reaches the specified weight, sensor *e* (also located on the scale) is actuated. This closes valve *A*, and after a short time delay to allow for the slight dripping action, the container is ejected by the action of cylinder *B*.

8-9-2 The Selection of Technologies (Fig. 8-16)

Sensors *d* and *e*, located on the scale, are of the interruptible jet type (see Section 4-3-4-3). This type of sensor is employed so that the accuracy of the scale is not disturbed by a contacting type of sensor.

Sensor *f* detects the presence of empty containers. In this application it can be a movable-part, mechanically actuated limit valve, or back-pressure sensor. The choice is determined by the weight of the empty container.

Cylinder *A* actuates a limit valve when the rod is retracted. No limit valve is required to sense the extended position because of sensor *e* located on the scale, which initiates the retraction of cylinder *A*.

When cylinder *B* and *C* are retracted, they both actuate limit valves. No limit valves are required to sense the extended position of these cylinders because the retraction of the cylinders is controlled by sensor *d* on the scale.

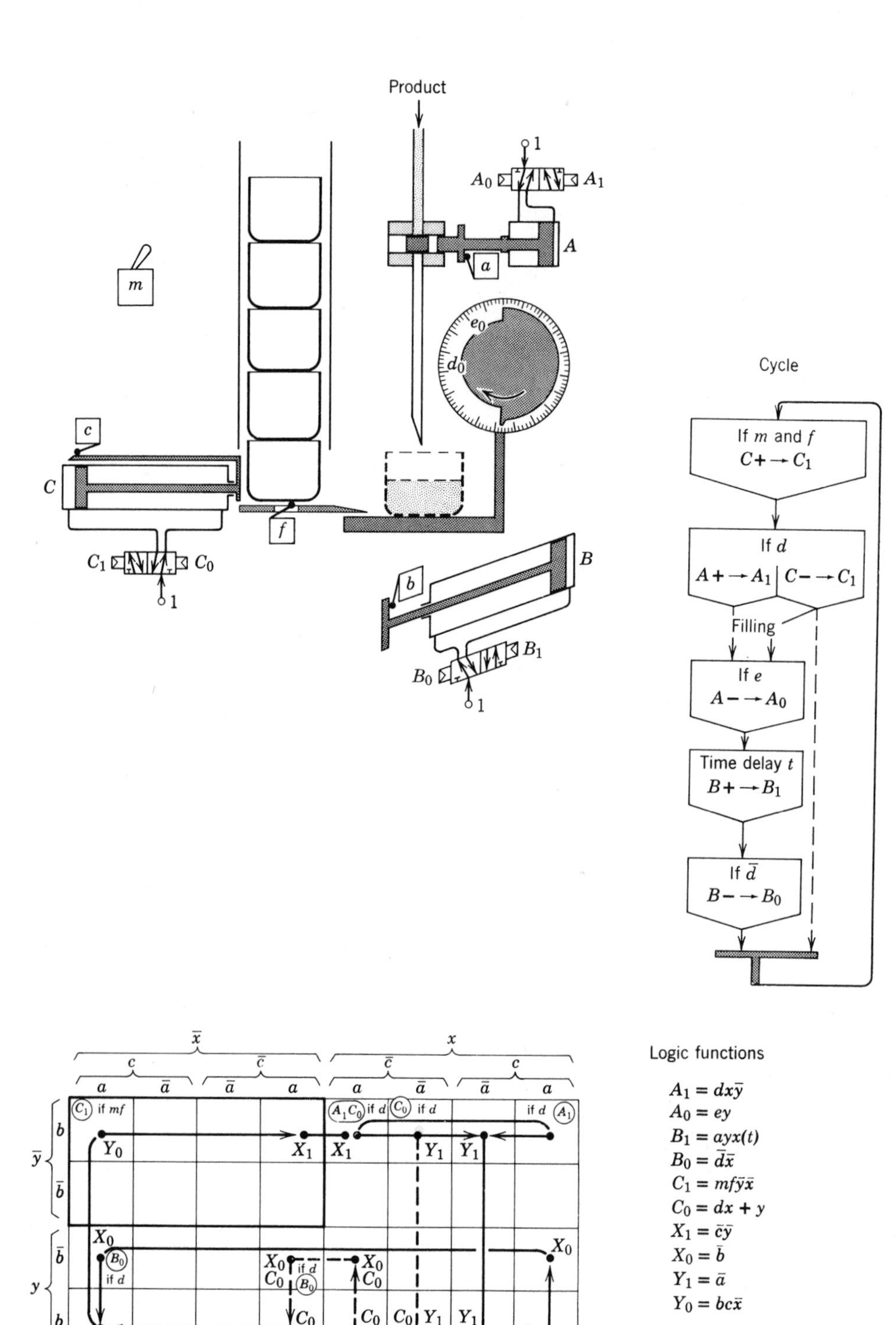

FIG. 8–16. Container filling method: typical cycle, Karnaugh map, and logic functions.

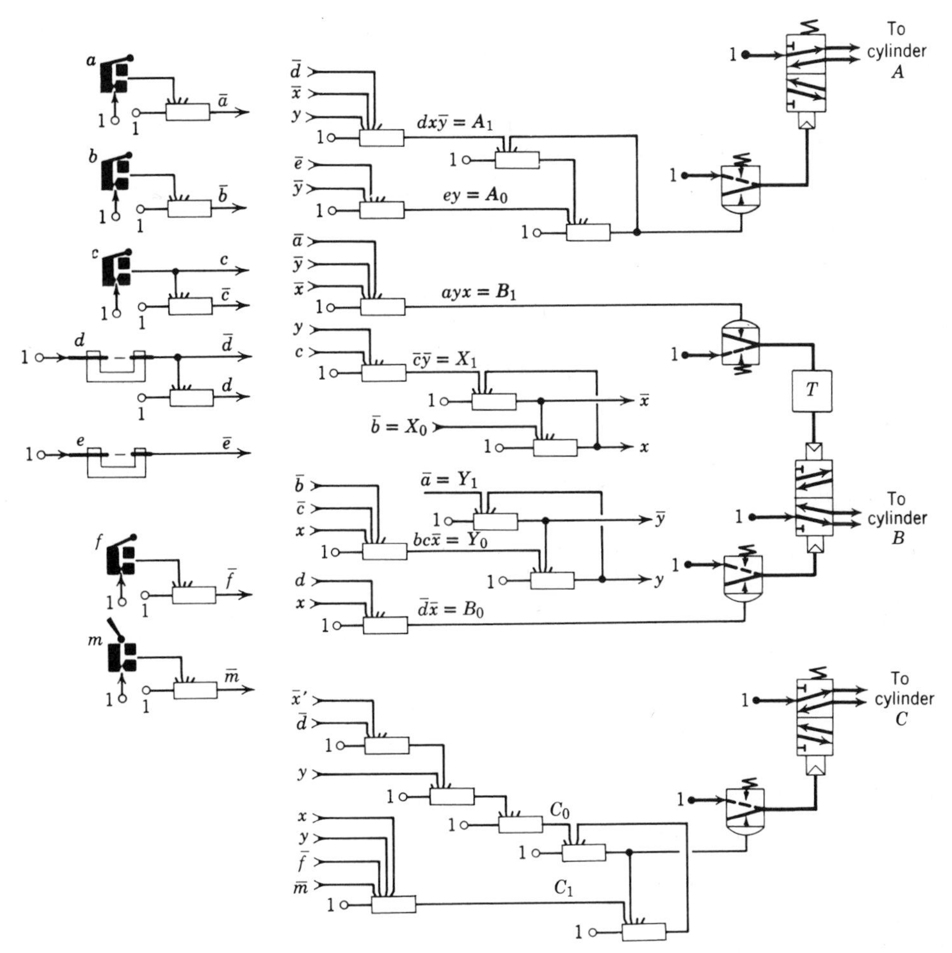

FIG. 8–17. Container filling logic circuit employing laminar/turbulent flow devices.

As a result of the utilization of both high- and low-pressure sensors, a comparison between the two types of circuits is facilitated.

8-9-3 Details and Circuitry

The Karnaugh map illustrates the need for two memory devices, X and Y. A possible cycle deviation is illustrated in Fig. 8-16 by the dotted line. If cylinder C is late in its retraction, the normal filling process must still be accomplished. During the deviated portion of the cycle, C_0 is maintained, for the retraction of cylinder C and the normal cycle can be continued at any point.

The circuit in Fig. 8-17 is designed with laminar/turbulent flow devices, employing a low-pressure supply and directly utilizing the low-pressure output signals from the nonmovable-part sensors. Output amplifiers pilot the power valves which actuate the cylinders. The required time delay is provided with high-pressure devices.

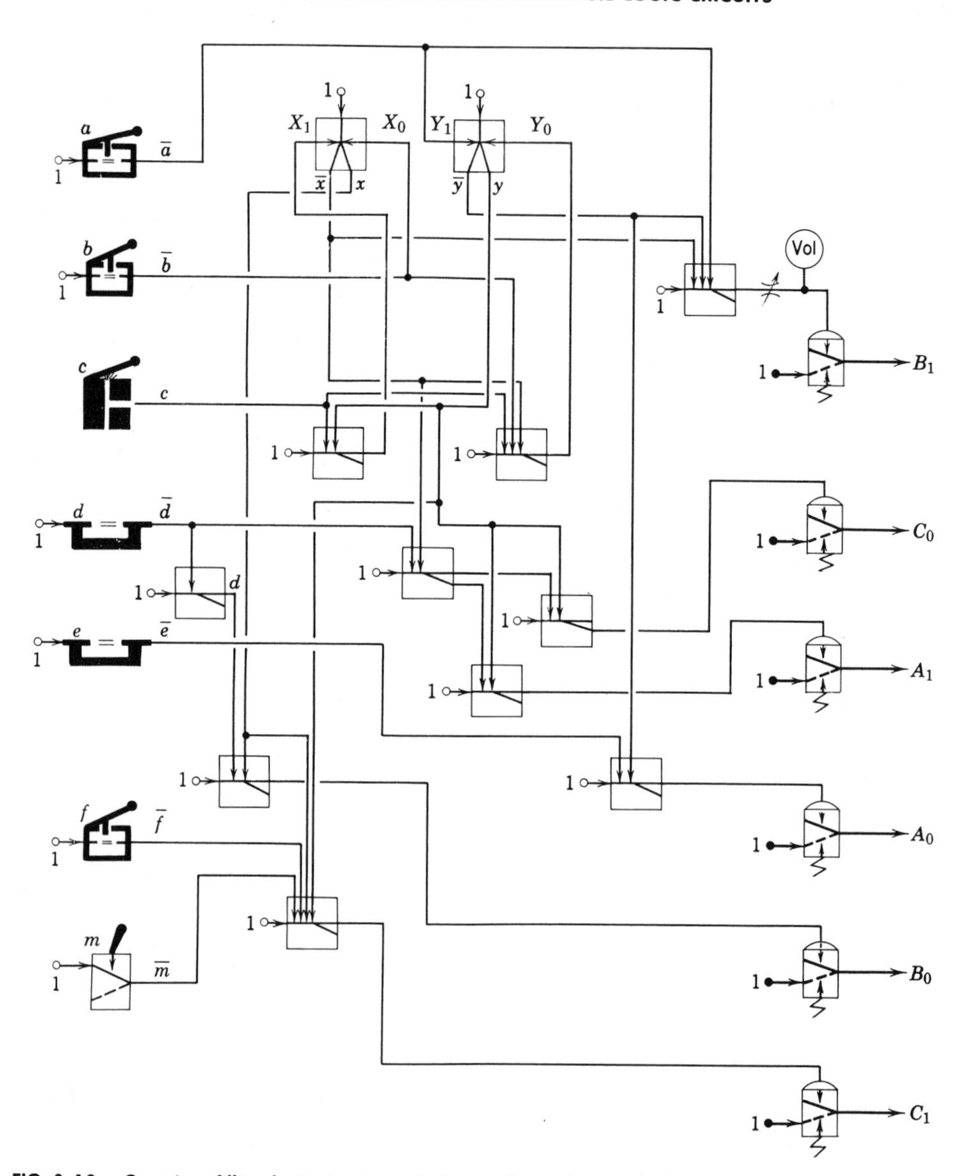

FIG. 8-18. Container filling logic circuit employing wall attachment devices.

Figure 8-18 illustrates a circuit designed with wall-attachment logic devices. This system is similar to the one above in that the low-pressure power supply is compatible with the logic and the sensors. Output amplifiers are also required in this circuit. The main difference is that the time delay is operated on the same low supply as the other devices in the system.

The circuit in Fig. 8-19, is designed with high-pressure movable-part logic devices. The outputs of the interruptible jet sensors require amplifiers to provide them with the required level of input pressure to the logic. No output amplifiers are required between the logic and the power valves.

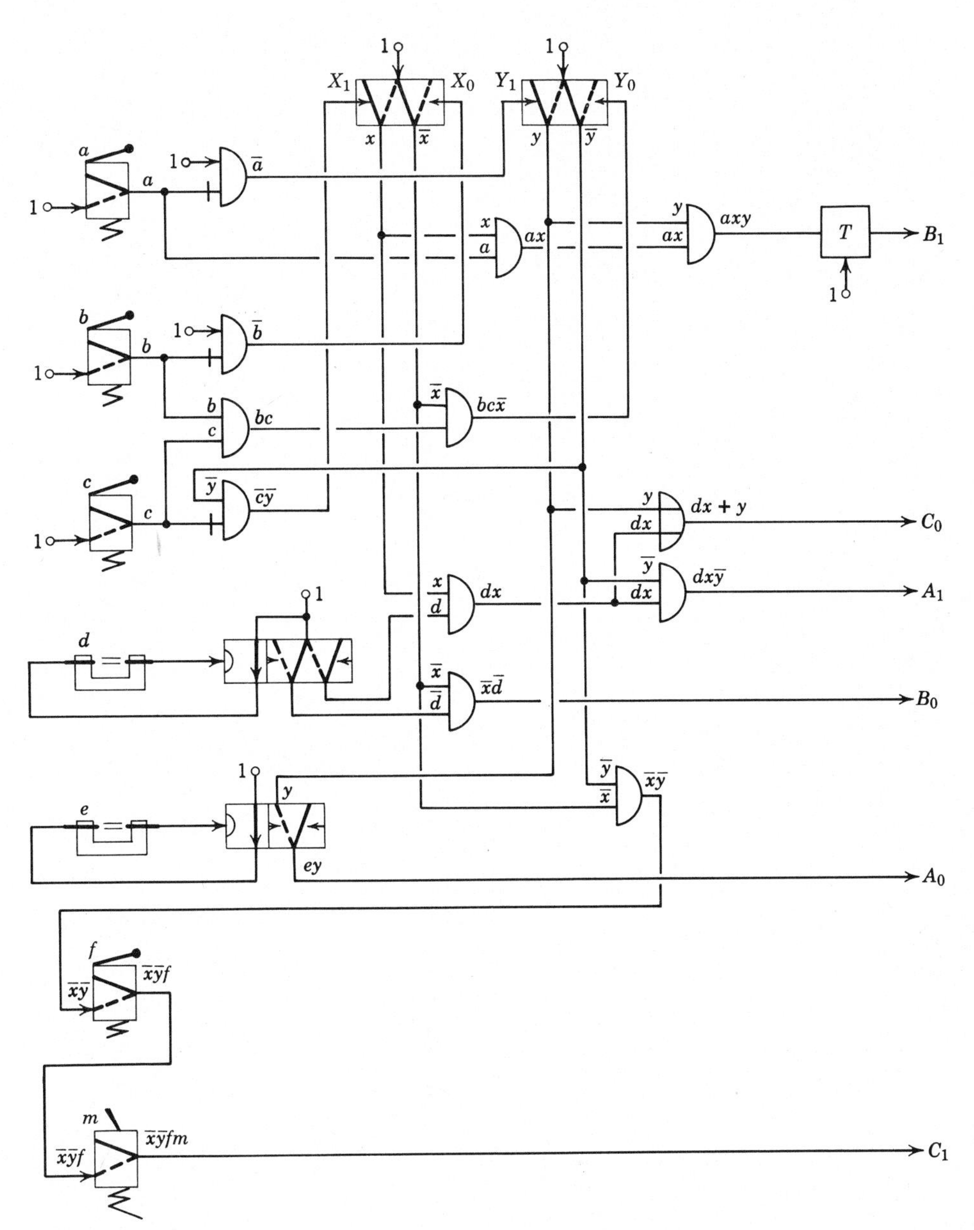

FIG. 8–19. Container filling logic circuit employing high-pressure movable-part logic devices.

8-10 MOLDING MACHINE

8-10-1 Description and Operating Principles

The machine in Fig. 8-20 molds the top and bottom halves of sand molds to be employed in the casting of metal objects. It is interesting to note the variety of technologies that must be employed in this machine: pneumatics, hydraulics, and electrics.

The sand is carried by a conveyor and drops into a hopper. The top of the hopper is then closed and air pressure is applied to the sand in the hopper to ensure that it fills all areas of the mold. A hydraulic cylinder compacts the mold and ejects it from the machine.

8-10-2 The Selection of Technologies

A hydraulic cylinder F is required because of the high level of force required.

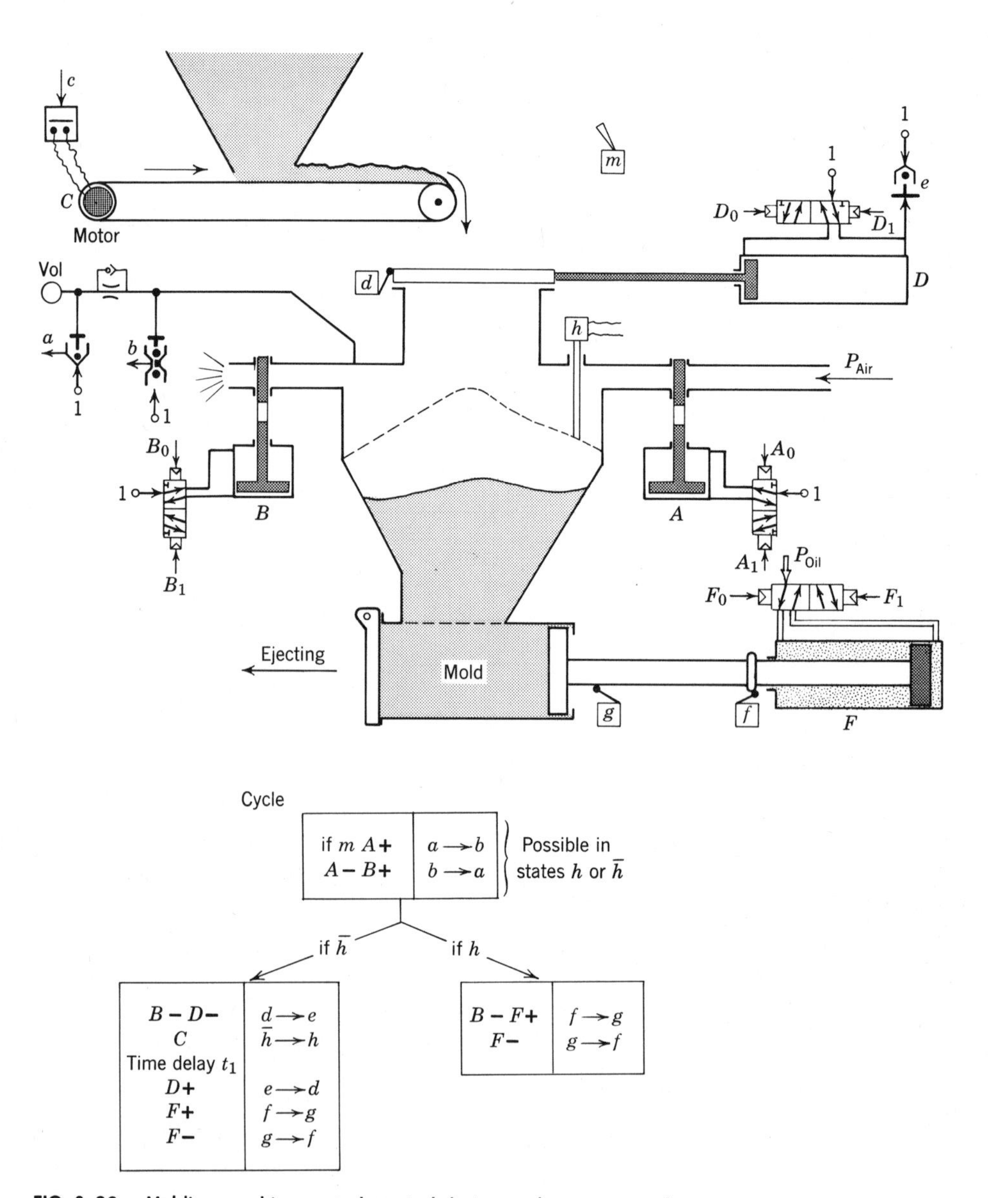

FIG. 8–20. Molding machine control; typical design and sequence cycle.

The use of hydraulic pressures up to 3000 psig or beyond is practical if required. A hydraulic valve with pneumatic pilots will provide the means to control the circuit with pneumatic logic controls.

Two pneumatic cylinders A and B provide the means to rapidly pressurize and exhaust the hopper. Cylinder A controls the supply pressure port and cylinder B controls the exhaust port.

Instead of employing limit valves for cylinders A and B, it is simpler and safer to detect the pressure in the hopper with pressure sensors a and b.

- Sensor a is a pressure release limit valve which senses when the hopper is at pressure 0. a signal is slightly delayed to ensure that the pressure is truly at zero.
- Sensor b is a pressure switch which senses when the hopper is pressurized.

The conveyor is run by an electric motor c which is controlled by a pressure switch actuated by pneumatic signal c. A level sensor h senses the required level of sand. This sensor must be electrical so that it will not be affected by the pressure variations in the hopper. Since the sand is moist, the electricity will have a path when the sand level reaches the sensor.

8-10-3 Details and Circuit

The cycle in Fig. 8-20 is as follows:

- When the starting switch m is actuated, the hopper is pressurized ($A+$ action) to compact the sand.
- When the required pressure level is reached, pressure sensor b initiates the opening of the exhaust valve on the hopper ($A-$, $B+$). Both of these operations can transpire, no matter what level signal is present in the hopper. Also during this period, the switching from state h to $\bar{h}$ can transpire but this action is not certain.

Depending on whether signal h is still present or not, the following two cycles can be initiated:

- If h is still present, there is no need to refill with sand and the cycle is shortened as follows:
  - The hopper exhaust port is closed ($B-$) and the mold is compacted and ejected ($F+$).
  - Cylinder F retracts ($F-$).
- If h is absent (state $\bar{h}$) refilling of the sand is necessary and the cycle is as follows:
  - Hopper exhaust port is closed ($E-$) and the hopper cover is opened ($D-$).
  - The motor is run until the sand reaches the level sensed by h.
  - Signal h stops motor C and after a short time delay, t, the hopper door closes, ($D+$).
  - The mold is compressed and ejected ($F+$).
  - Cylinder F retracts ($F-$).

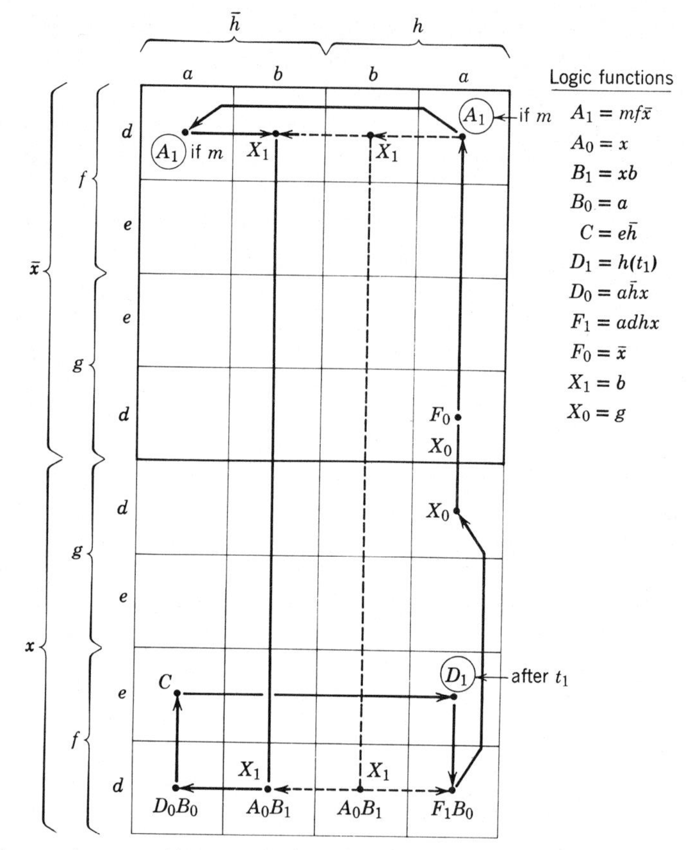

FIG. 8-21. Karnaugh map and logic functions for molding machine control.

At the completion of this action a switch from state h to state $\bar{h}$ is possible.

On the Karnaugh map in Fig. 8-21, the two possible cycles are drawn. The cycle start is possible in state $\bar{h}$ (solid line) or in state h (dotted line). At each step in the dotted line cycle, a passage from state h to state $\bar{h}$ is illustrated to utilize the full line cycle.

The cycle completion, that is, the compacting and ejecting of the mold and the retracting of cylinder F, is common to both cycles.

The logic equations are simple: the circuit in Fig. 8-22 is designed easily with high-pressure logic devices.

The electrical devices require interfaces:

- Solenoid valve h.
- A pressure to electric switch to control electric motor C.

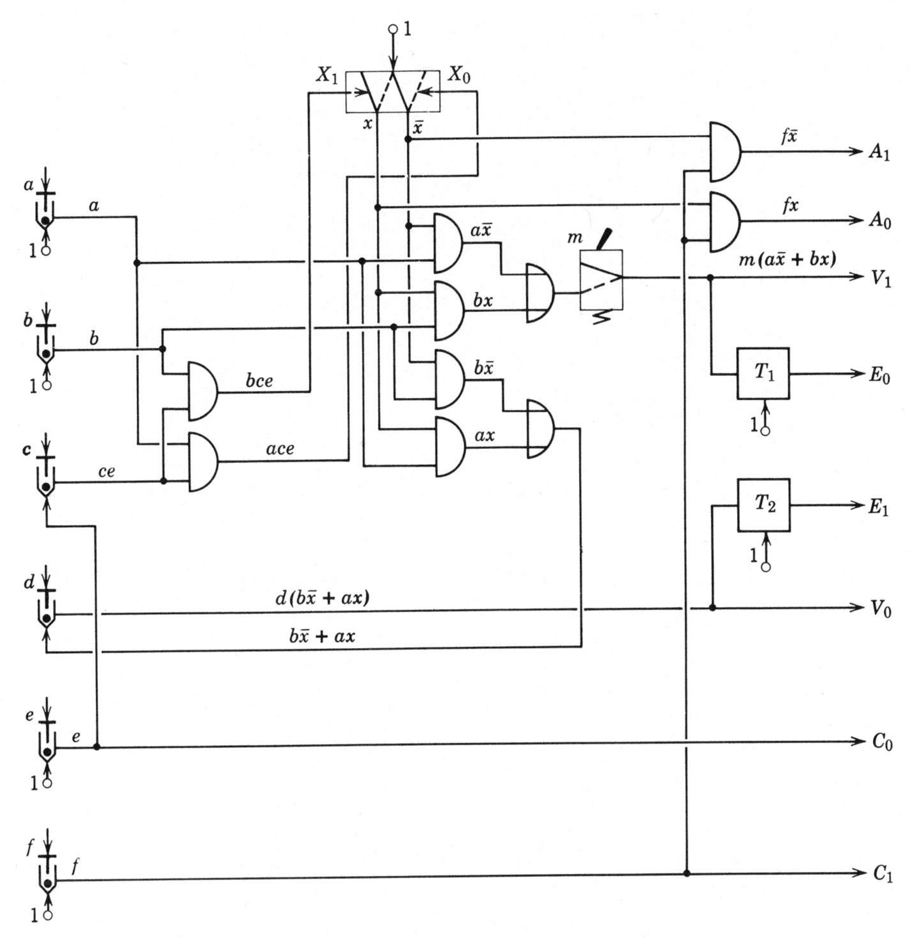

FIG. 8-22. Molding machine control—high-pressure movable-part logic circuit.

8-11 PRODUCT LOADING FOR HEAT TREATMENT

8-11-1 Description and Operating Principles

Products to be treated are lined up one by one in rows of 2, 3, 4, . . . to be fed into treatment facility (oven or drying room), or possibly a spraying booth or packaging machine. This type of product handling is required in many types of industries.

Figure 8-23 illustrates a system which lines up products in rows of five before feeding them into the oven.

Cylinder *A* positions products one by one into locations *F, E, D, C,* and *B*. Cylinder *I* controls the oven door and cylinder *G* pushes the row of five products into the oven.

Dairy tank flush controller. (Courtesy Industrial Automation Works, Czechoslovakia.) The cabinet in the foreground above houses power supplies, regulators, filters, and logic programming hardward to control the cleaning cycle of milk tanks in the background. Indicators on the front of the cabinet announce sequence progress as well as identify the controllers in operation. The logic hardware of this system, called PNEULOG, consists of trays of low pressure diaphragm devices shown below. Six basic modules fulfill the popular logic functions in this system. Power for each module flows from the tray and all signaling is piped from the tops of elements. The trays slide into racks and are interconnected from their end fittings.

8-11-2 The Selection of Technologies

Sensor k indicates the products presence in front of cylinder A. The sensor must be selected according to the type of object that is to be sensed. A no-movable-part sensor is recommended for objects that are light weight and

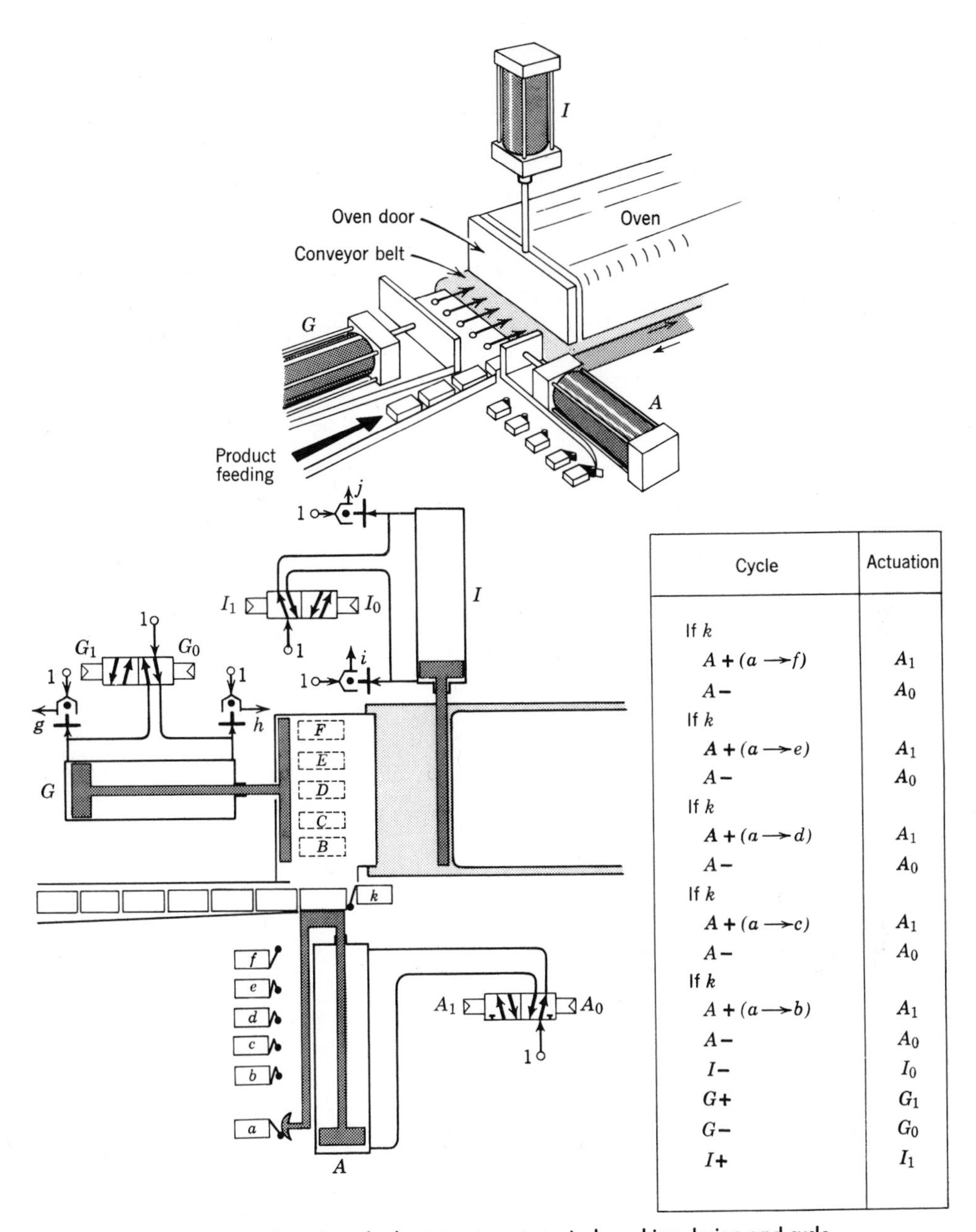

Cycle	Actuation
If k	
$A+(a \rightarrow f)$	A_1
$A-$	A_0
If k	
$A+(a \rightarrow e)$	A_1
$A-$	A_0
If k	
$A+(a \rightarrow d)$	A_1
$A-$	A_0
If k	
$A+(a \rightarrow c)$	A_1
$A-$	A_0
If k	
$A+(a \rightarrow b)$	A_1
$A-$	A_0
$I-$	I_0
$G+$	G_1
$G-$	G_0
$I+$	I_1

FIG. 8–23. The loading of products for heat treatment; typical machine design and cycle.

unusually shaped (interruptible jet or back-pressure sensors are typical). For heavier objects a mechanically actuated limit valve is practical.

Limit valves *b, c, d, e,* and *f* control the rod extention of cylinder *A*. Since they are needed for only one direction of rod positioning, limit valves with one-way rollers are recommended to simplify the circuit. Because limit valve *f* is at the end of the row it will operate satisfactorily with a straight roller type. Pressure release limit valves employed in conjunction with cylinder G and I simplify the machine and provide reliable feedback information.

8-11-3 Details and Circuit

No difficulties are encountered as a result of employing one-way roller-type limit valves. The number of memories has been limited to three. Figure 8-24 illustrates the Karnaugh map and the logic functions. Fig. 8-25 illustrates the high-pressure movable-part logic circuit.

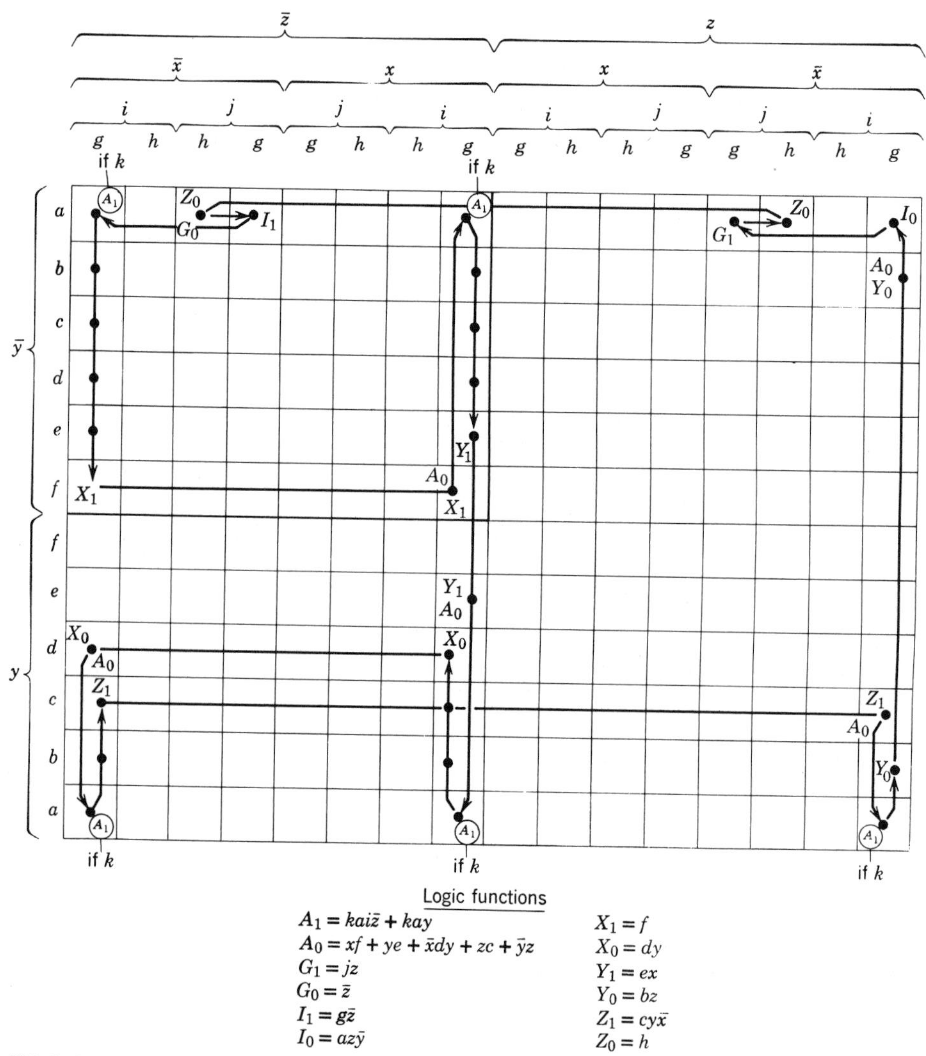

FIG. 8-24. Logic functions and Karnaugh map for product loader.

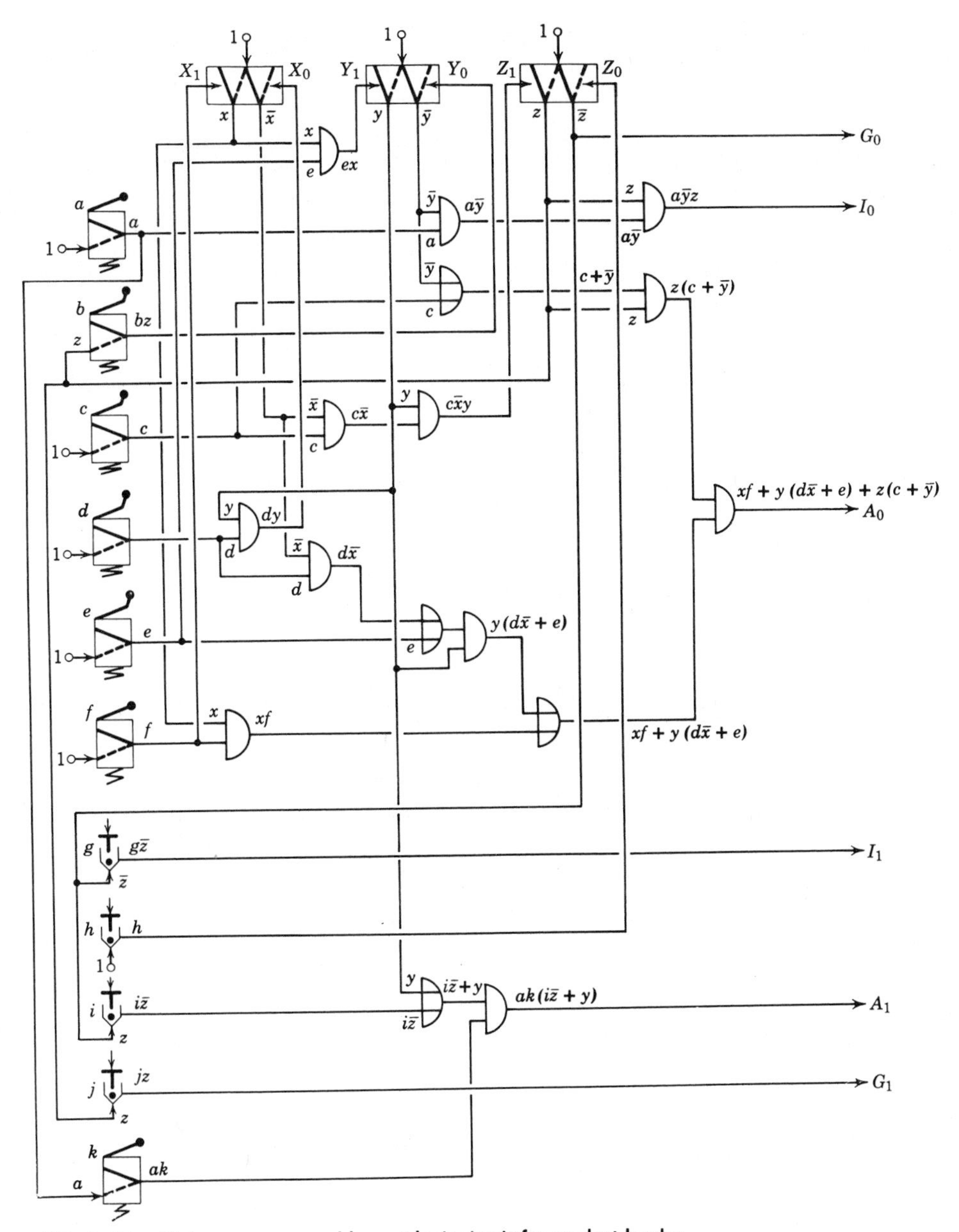

FIG. 8–25. High-pressure movable-part logic circuit for product loader.

8-11-4 Alternative Circuit Designs with Various Types of Logic Devices

The sensing of positions *F, E, D, C,* and *B* can be accomplished by the products themselves directly actuating the sensors.

This new design is illustrated in Fig. 8-26.

Sensors *b, c, d, e, f,* and *k* depend on the weight and shape of the product. Typically back-pressure sensors have been selected.

A compatible circuit utilizes wall-attachment logic devices with the sensor. The devices operate at pressures below the levels required by pressure release limit valves. Cylinders *G* and *I* are equipped with mechanically operated limit valves.

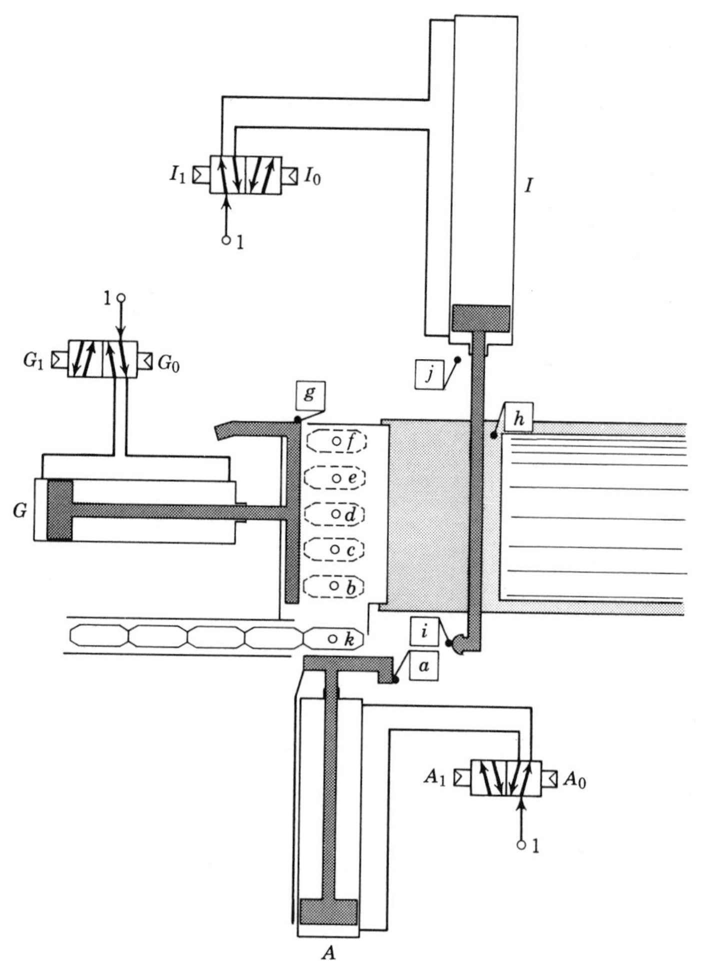

FIG. 8–26. Typical design of product loader employing back-pressure sensors.

The details of this alternative solution is illustrated in Fig. 8-27. A few points worth noting:

- Variables *a, b, c, d, e,* and *f* are now compatible—a full size map is required.
- Sensors *b, c, d,* and *e* are actuated and deactuated during the positioning of each product. This creates many actions in the states $\bar{a}$, $\bar{b}$, $\bar{c}$, $\bar{d}$, $\bar{e}$, and $\bar{f}$ which corresponds to no product being sensed. An example of this is the positioning of the first product at sensor *f*. Several paths through a single square are possible because no action results until the combination is right.

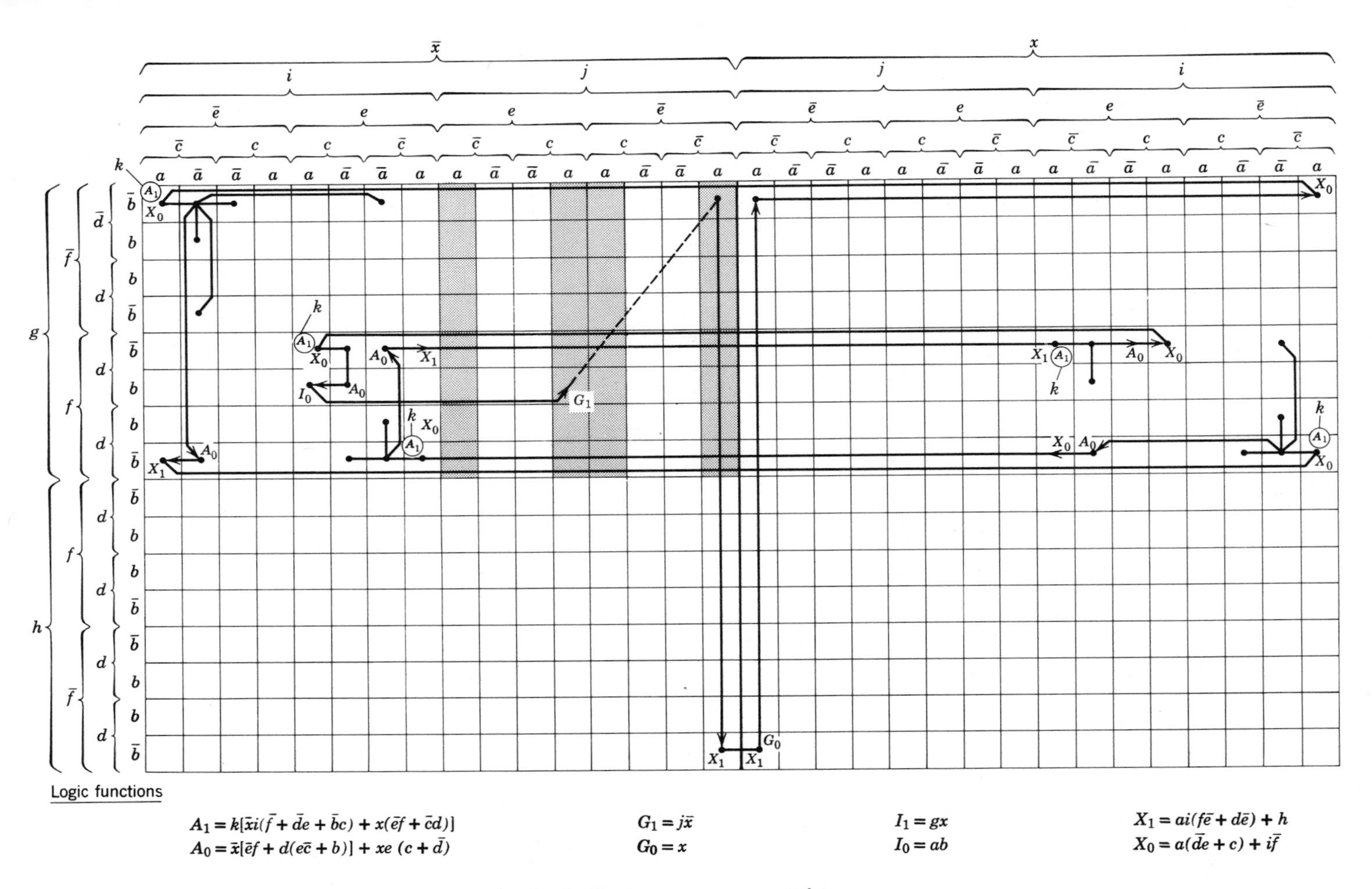

Logic functions

$$A_1 = k[\bar{x}i(\bar{f} + \bar{d}e + \bar{b}c) + x(\bar{e}\bar{f} + \bar{c}d)] \qquad A_0 = \bar{x}[\bar{e}f + d(e\bar{c} + b)] + xe\,(c + \bar{d})$$

$$G_1 = j\bar{x} \qquad G_0 = x$$

$$I_1 = gx \qquad I_0 = ab$$

$$X_1 = ai(f\bar{e} + d\bar{e}) + h \qquad X_0 = a(\bar{d}e + c) + i\bar{f}$$

FIG. 8–27. Karnaugh map and logic functions of product loader (back-pressure sensors style).

- When cylinder G is actuated all sensors (b, c, d, e, f) are deactuated in a miscellaneous order. The changes of state for all of these sensors transpire while g is still actuated, as a result of its long cam. The shaded squares of the Karnaugh map illustrates all of the states that can be encountered during the operation.

It can also be noted that since the products differentiate their various states by the actuation of their individual sensors, the number of memory functions is greatly reduced. Figure 8-28 illustrates the solution, employing-wall attachement devices.

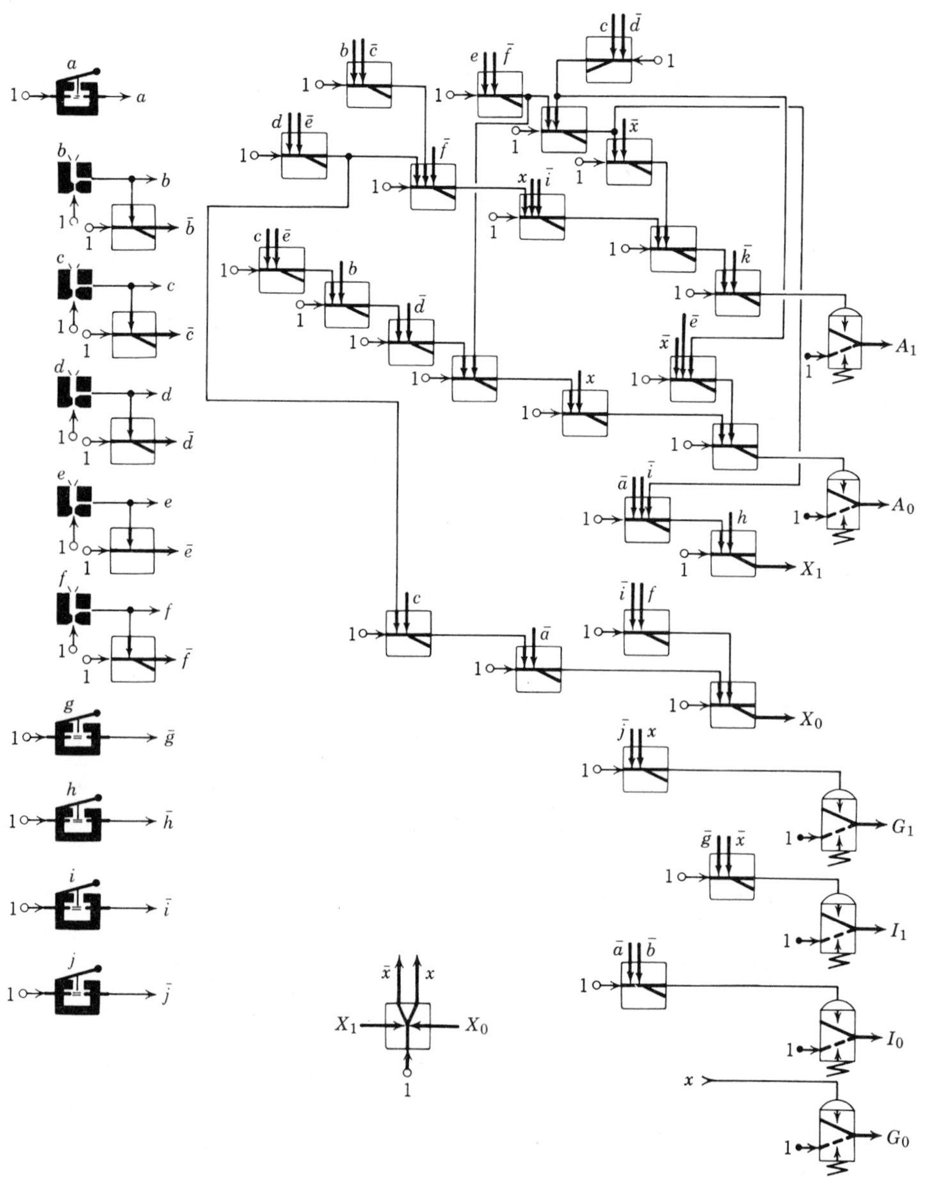

FIG. 8-28. Product loader circuit employing wall attachment devices.

CHAPTER 9

THE FUTURE OF FLUID POWER AND FLUID LOGIC IN INDUSTRIAL AUTOMATION

This chapter attempts to give to the reader an overview of the developments that are influencing the future course of industrial automation and its relationship to fluid logic and fluid power systems.

Chapter 1 and 5 discussed in detail the advantages of machines that are controlled and powered by devices utilizing fluids as the energy media. The application of fluid operated devices throughout a machine, from the sensors to the power systems, maximizes their numerous advantages, among which simplicity of design and reliability of operation are primary.

9-1 THE INCREASED UTILIZATION OF FLUID POWER SYSTEMS

As discussed in Chapter 1, the need for linear motion in industrial automation led to the development of pneumatic and hydraulic cylinders. It is our firm belief that the continual progress in the application of fluid logic controls will increase the pace at which fluid power systems are being accepted by industries worldwide.

In some applications electromagnetic actuators have been retained because of their homogeneity with the electrical systems that controlled them.

As fluid logic control systems replace the electric controls, it becomes more advantageous to replace also the electrical actuators with cylinders, that are inherently more reliable, compact, and more economical.

The traditional utilization of cams, gears, and screws to create linear motion is declining at an ever-increasing rate. Frequently these power systems are replaced by fluid power systems with electrical controls. From this interim stage it is a simple step to go to a totally fluid operated system. The elimination of mechanical power systems greatly simplifies the design and construction of machines, and also facilitates any future modifications to the machines, because of the elimination of special devices, for example, cams, levers, and gears.

To date, cylinders have been the primary fluid power actuators in industrial automation. However, many other types of fluid power devices are being utilized to simplify the automation process:

- Air jets are the simplest fluid power devices. They are used to clean tooling between cycles, to eject finished parts, and to cool molds. Pressure and flow are easily controlled to meet all requirements, and air consumption is minimized by short cycle times.
- Venturi devices provide a means of creating partial vacuum (50 to 70%). Vacuum pads for material handling is a common application which can be easily integrated into pneumatic control systems. This application is illustrated by the example in section 8-8. Objects weighing 1 oz. to 1 ton can be easily lifted. This type of system is equivalent to electromagnets. However, these are limited to the handling of ferrous materials, in contrast to vacuum pads which can be used on ferrous and nonferrous materials.
- The combination of an air jet and a venturi creates a spray gun. The spraying of paint and other chemicals is universal and also easily integrated into fluid logic control systems.
- Inflatable tubes or air bags can perform mechanical functions:
    - The holding or lifting of parts with air bags.
    - Displacement of parts along a tube compressed by rollers.
    - Fluid flow valves made of inflatable tubes.
    - Inflatable door seals, and such.

    - Fluid powered motors:

    - Pneumatic turbines can operate at a very high speed. However, their torque ratio decreases as speed is reduced. They are commonly applied to industry for automation drilling, grinding, and tapping. A wide variety of portable hand tools are also employed on assembly lines and on site construction jobs. Overall air-operated tools offer more power safety, compactness, and ruggedness than equivalent electrical tools.
    - Hydraulic piston motors in contrast to air motors provide low controllable speeds with very high torques.

The list above is far from complete. We have not mentioned some specialized devices that are peculiar to particular industries. We have only attempted to suggest the possibilities that are available now. From these we can project a very bright future for fluid power devices and their place in the automation of industry.

9-2 THE INCREASING UTILIZATION OF FLUID LOGIC CONTROLS

Fluid power and fluid logic controls are progressing along parallel paths, each one stimulating the use of the other.

- Fluid power provides physical actions.
- Fluid logic provides sensing and programming.

Although we have not discussed the utilization of fluid logic controls in hydraulic power systems, it is important to remember that hydraulic power and pneumatic power systems are equal partners in the technology of fluid power and the value of both should not be overlooked if the potential of fluid logic control is to be maximized now and in the future.

9-3 'THINK FLUID'

"Think Fluid" is the key phrase that must be protected if the potential of fluids as power and control media are to be realized. It is up to the engineers of the world to take advantage of this important technology *now* if it is to remain a viable aid today and become indispensible to industrial automation tomorrow.

APPENDIX

GRAPHIC SYMBOLOGY

The subject of graphic symbology for fluid logic circuitry has become as controversial as the technology it represents. Most manufacturers of fluid logic hardware feel that their system is so unique as to warrant a special symbology of its own. This resulting plethora of graphic art has confused and irritated the users to the point that many systems designers have synthesized and modified the various symbols to create a method that is comfortable and readily understandable for their own purpose.

As the situation stands today, systems of graphic symbology are widely employed and have received recognition from various organizations concerned with standards. Fig. A-1 represents four symbologies, each being inspired by a given technique. Collectively as author and editors we do not endorse any of them because like every one else, each of us has his personal preferences.

It is our unanimous hope that in the near future, as the technology of fluid logic and other logic technologies develop around the world, representatives of the industrialized countries will meet to resolve the problems of graphic symbology. We optimistically look forward to the day when international standards will be adopted and accepted by system designers in all countries of the world.

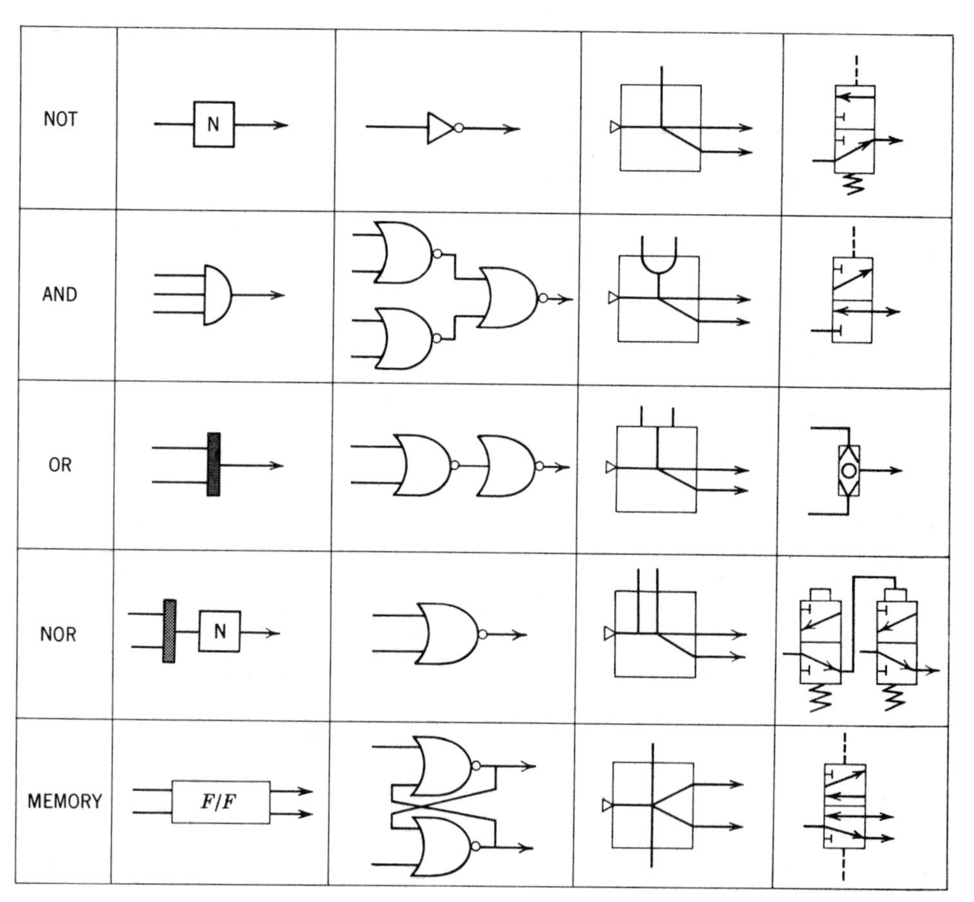

FIG. A–1. Typical logic symbols in use in the United States.

INDEX